Best wishes, hope you will en[illegible] the Book

Robert L. Casey

AROUND AND AROUND HE GOES

by

Robert L. Casey

FIRST EDITION

Library of Congress Catalog Card No: 98-90501
ISBN: 1-56002-841-6

UNIVERSITY EDITIONS, Inc.
1905 Madison Avenue
Huntington, West Virginia 25704

CONTENTS

FOREWORD

These writings cover the life and times of Robert Lee Casey. The doctor who brought him into the world was an old family friend. Robert's parents named him Robert Lee, but the doctor put the name Mike, Jr. on the birth certificate. The family was not aware of this until 1939 when Bob received the alternates appointment to West Point. When they applied for a birth certificate, the name shown was Mike Casey, Jr. Since everything to date was questionable, steps were taken to change the name to Robert Lee Casey.

This book is about 95% true. All of the more important events and comments can be authenticated by pictures, newspaper clippings, letters, general orders, etc. Most of the people in this book are real people. Many of the names are not known or have been forgotten over the years. It is not the intention of the author to discredit or embarrass anyone. Some of the dates may be a little off the mark. The writer did very little in the way of research, but he wrote it all from memory. The bulk of the manuscript was written by hand, and it was quite a chore. The author is a victim of Parkinson's disease. This drastically affected the writing in a non-beneficial way. Many significant events took place during the lifetime of Bob Casey. He participated in many of these happenings and in a modest way made his contributions.

Mike Casey, Jr. was the name selected as the central character of the book. He is not meant to appear a hero. Many of his disappointments and failures are mentioned in the book - his performance in high school, his color blindness, his failure of the West Point examination, his failure to be promoted during the three years of his enlistment June 1946 to June 1949, and his first semester at Marshall University.

Robert (Bob) successfully engaged in three careers with a modicum of success. He retired from the West Virginia State Police in 1965 as a District Commander. He occupied a full colonel mobilization assignment as a military counterpart of the State Director of Civil Defense. He retired in 1981 with the permanent rank of lieutenant colonel. When Bob was assigned to Israel as Chief Security and Safety Officer, he was promoted to GS-13, a very respectable grade for a Huntington District employee. In the Sinai he was assigned to a State Department project that was critical to the peace accords. If very many people read this book, Bob may add "author" as another career.

Bob kept himself in good physical condition. During his middle years, he jogged three or four miles a day. He grew up in a home that was headed by a handicapped father and assisted by a truly wonderful mother, who was the chief of staff. Sadie was a strong-willed woman who was fiercely protective of her boys. She tolerated no foul language or questionable literature in her home. Mike was taught to respect women, open doors for them, let them enter a room first, and say "Yes, ma'am" to older ladies, and give up his seat to them. Sadie taught a Sunday School class and never missed taking her boys to church. Bob caused her a lot of grief with his wanderings, but he always loved and respected her. Bob's father served as a scoutmaster for several years and later became a scout commissioner. He helped organize the Beverly Hills Boys' Club. Mike, Sr. served 12 terms in the West Virginia State Legislature. He was an avid supporter of Marshall College and helped them to attain university status.

Bob was greatly influenced by his father. Mike, Sr. passed away in 1979. Seldom does a day pass that Bob does not think of him. Bob hopes that Mike, Sr. has a ringside seat and is reasonably well-pleased with the activities of his son.

Martha and Bob have two daughters, both are intelligent and talented. The older, Saundra, retired from the U.S. Air Force with the rank of Major and now heads up a JROTC program near Savannah, Georgia. Sandy has a doctorate in Adult Education from Auburn University. Martha, the younger daughter, graduated from Indiana University with High Distinction with a Bachelor of Science in Education and earned a Master of Arts with a major in Special Education from Marshall University. She currently is a middle school teacher.

Bob started at Marshall College in 1939 and finished in 1975. He has six hours toward a Master's Degree.

ACKNOWLEDGEMENTS

I recognize and acknowledge the support received from my spouse during these trying times. Martha provided me with medication at the proper intervals and in the right amounts. She encouraged me to write a little each day and get plenty of rest. If there is anything more demoralizing and uncomfortable and heart-breaking than having Parkinson disease; being a caregiver must rank right near the top.

My daughter, Dr. Saundra L. Casey deserves an accolade for her contribution. She de-ciphered and typed the almost, illegible, hand written manuscript. This was all accomplished long distance, as her ROTC assignment was near Savannah, Ga. This must have been a labor of love and faith as she had no assurance the project would ever be completed.

CHAPTER 1

Around and Around He Goes

Inside the confines of a small, tacky kitchen a blue fly droned aimlessly. The kitchen was located on the top floor of a drugstore in a small, isolated community. Suddenly a young woman, obviously heavy with child, came lumbering into the room. She attempted to sit on one of the frail kitchen chairs, but missed and landed on the floor. At that precise moment a lusty son was born. Mike and Sadie Casey had made their first contribution toward helping populate the universe. From the vigorous sounds and movements this one promised to be a dandy. Using pre-arranged signals, the father who was working at a nearby railroad yard was summoned and he raced to the apartment adding more noise and confusion to the scene.

The doctor arrived and everything was under control. The baby was named Mike, Jr. This writer would never have known the lowly circumstances of his birth if he had kept his mouth shut thirty years later and allowed his spouse, Martha, the last word in a family argument. One afternoon they were discussing different size families. Martha observed that she had five brothers and sisters and he had only one brother. His immediate response was "scrub stock always comes in litters." It was a deadly, devastating mistake. From that day on any critical or uncomplimentary remark was always answered by, "Well! I wasn't born on a kitchen floor." Mike, Jr. was born hungry, but he quickly solved this problem and went on to bigger things.

Exactly twenty years and two months later Mike, Jr. was crossing the US-Canada border by train enroute to Ottawa. He was still hungry, but this time for adventure and excitement. This along with a subconscious desire to make the world a better place was his motivation. A fierce air war was being waged by the Germans against Great Britain. The British army had just suffered a humiliating defeat at Dunkirk. They were tired, hungry, and frustrated. They were short on ammunition, equipment, and food. They were not getting the assistance they had expected from the US. British convoys were being decimated by the German Wolf Packs. Morale was low. During a one month period over 60% of the supplies and equipment, leaving from ports all over the world, were destroyed. The British people were spending their nights in subways and other underground shelters, hoping to survive the unforgettable experience of being blitzed.

Mike was aware of the precarious position of the British and how it could affect not only England but the entire world. He and his grandmother had spent many evenings discussing the war in Europe and its potential for disaster on a grand scale.

In the spring of 1941, Mike enrolled in the National Youth Administration (NYA) and received private tutoring in math. Cadets for pilot training were in great demand and West Virginia was not producing its fair share. The extra effort resulted in his passing the mental requirements for appointment. No one had ever considered that a healthy youth like Mike could fail a physical exam, but lightening struck and Mike experienced the most bitter disappointment of his entire life. He was declared color-blind and could not qualify for the cadet program. Mike and a friend, who had failed a depth perception test, decided to go to Canada and enlist in the RCAF. Hasty calculations indicated that it would require about $50.00 each to get to Canada and subsist for a few days. They decided to meet at the Charleston Bus Station the next day. Mike financed his trip entirely with funds he had stashed away in his room. They rode a bus to New York. Upon arrival at New York, imagine their delight to see a RCAF cadet leaving the station. He was due to graduate in two weeks and was going home on leave. The two would-be aces arrived at a RCAF recruiting facility the next day and started processing. Mike was denied enlistment because of color blindness, but his friend was accepted. Instead of accepting defeat and going home, Mike decided to join the Royal Canadian Ground Forces. Mike enlisted in a Canadian Scottish unit named "The Stormont Dundas Glengarry Highlanders" and was issued a World War I dishpan helmet and an old Lee Ensfield Rifle.

Eleven days after his enlistment Mike was on a ship headed for England and a very uncertain future. The trip from Ottawa to Halifax, Nova Scotia was by rail. During the rail trip Mike participated in a card game which the Canadians called Pontoon. It was much like American Black Jack. After several hours Mike had most of the money. A short, swarthy French Canadian asked Mike to lend him $2.00 and promised to repay the money when he got to his luggage later on. Against his better judgment Mike let him have the money. About 45 minutes later, the Frenchman lost all of the two dollars and decided to quit.

Upon arrival at Halifax, the troops were loaded immediately and were led to a compartment which was supposed to house 160 men. The various barracks bags and other equipment had already been stacked at one end of the large room. Probably 50 men would have been enough to create a shortage of space. The compartment was filled with tables. Each table had about a one inch rim around the edge. Later Mike learned that the rim was to keep the mess kits on the table in rough seas. There were small hooks welded to the steel ceiling. He later learned that those hooks were provided to support large, dirty hammocks. There were also thin lumpy mattresses if a soldier wanted to sleep on top of the tables. Some slept on the floor under the tables. So! Three layers of tired disgusted troops were provided space in the now unpleasant and smelly place below decks. Mike could visualize what would happen if a torpedo struck the ship. It would create utter confusion and the fatality rate would be enormous. There was no fresh water for showers and the salt water which was available was not heated. A salt water shower is no bargain hot or cold. Most of the men took no shower at all. The troops were permitted to place their hammocks on deck if they could locate a place with hooks to attach them to. Mike immediately grabbed one of the much used and abused hammocks and headed topside to locate a place to lay his head that first night aboard ship. Mike soon learned that the hammocks could not be in place before 1800 hours. He tied a tag with his name and rank and serial number on the soon-to-be bed—Pte Mike Casey, C31637.

Rumors were quickly being spread, digested, and analyzed. The best guess was that the ship was headed for England. The convoy was made up of 15 or 20 ships, both armed and transport, and naval units to protect the convoy.

A ship's bell signaled that it was chow time. All of the troops were hungry and headed for the troop mess. The area where the soldiers were to sleep, eat, and play was the same smelly area they were led to when they first boarded ship. Fresh air was constantly fed into the room by a large fan, but the volume was not enough to do the job.

The meal consisted of Bully Beef (Australian horse meat) and a slice of bread. This and a mess kit cupful of tea made up the menu. Bitterly disappointed and still hungry, Mike went topside to make his hammock a bed. British battle dress was one of the most uncomfortable uniforms in the world. It was heavy, stiff, and itchy. Made up of heavy wool, it never fit unless altered in some manner by a tailor. Heavy brogans with horseshoe heels and wrap leggings made up the balance of the uniform.

The deck was so crowded with hammocks that Mike found it necessary to locate near the railings. He had been issued two blankets. Mike spread one blanket out on the hammock and folded the other to fit on top. He decided to sleep in his battle dress both for the warmth provided and the safety and convenience of being already clothed in case disaster struck. When Mike settled down and fell into a troubled sleep his left foot was sticking out over the cold, Atlantic Ocean.

Mike began to think things out and wondered why he had ever gotten himself in such a predicament. He thought about his mother and the abrupt way he left home without telling her. He had pulled this thoughtless caper once before when he went to Nevada with the CCC (Civilian Conservation Corps). His mother's hair was turning prematurely gray from concern and speculation as to what he might do next.

The first night at sea was uneventful as far as hostile or aggressive activities were concerned. Things soon changed as the destroyers practiced their patterns. The ominous sounds of ships clearing their weapons and dashing hither and yon at flank speed was very disconcerting. Several of the transports cleared their weapons and readied themselves for the inevitable attack.

Breakfast was served at 0700 hours in the same greasy mess gear used the night before. The second meal served on the ship consisted of a half mess cupful of porridge (oatmeal). They did serve sugar with the porridge. The cooks prepared plenty of hot water and the mess gear was cleaner than the night before. Shortly after the mess gear was stowed. Life boat practice began and lasted until just before noon. Dinner was served and guess what? Bully Beef was served once more. They must have killed every horse in Australia by the war's end. The troops just loafed around and played Pontoon. The French Canadian who had borrowed the two dollars, came by and paid up. He didn't appear to be too happy. That night "Frenchie" disappeared and was never seen again. The ship was searched from stem to stern. Obviously Frenchie either jumped or was thrown overboard. Mike was not aware of any serious investigation of the incident or accident.

The convoy changed course every few minutes to confuse the submarines. That is why it took so much time for a crossing. There were three attempts to destroy the convoy. They could see and hear the destroyers dashing madly from one place to another dropping depth charges. The shrill bark of their sirens added to the gravity of the situation. The crossing took about one week and ended at Bristol Harbor which had been bombed incessantly. Many of the buildings and homes in the area had been hit with incendiary bombs, resulting in extensive repair work. Ships docked for unloading were hit and masts of vessels were sticking up out of the water. Gantry cranes used for handling cargo were heavily damaged. The weird, ridiculous barrage balloons were in evidence at strategic locations. The barrage balloons were strung with wire, dangling down from the frame. They were a definite hazard for low-flying aircraft. The other commonwealth troops used to tell the Brits that the balloons were the only things that kept the land mass of Great Britain from sinking into the ocean. Of course the harassed Brits resented it and many a fight resulted from it. The transfer from ship to train was accomplished and the troops were glad to get their land legs back.

Most of the time on the ship was spent just laying around and playing cards. Sgt. West had his men assembled once a day for some form of exercise. There were several Americans on the ship. Mike did not keep a diary and does not know the names involved. One of the men was a former New York state policeman. Another was a New York doctor who was given the rank of captain because of his profession. Two other Americans were from Massachusetts.

At last they learned where they were to be stationed. Aldershot, England. Upon arrival at the new billets the troops were assigned quarters, which consisted of an iron cot in a dismal brick and stone barracks. The iron cot was in two pieces and one end slid under the other, creating more space for other activity. The quarters were designed to provide just the basics necessary to exist. British private soldiers were never meant to be comfortable in their billets. No lockers or drawer space was provided. Chairs were nonexistent. Hot water was not available. About 50 square feet of floor space was provided to each private. There was no relaxation such as taking a nap on the bunk. It just wasn't comfortable.

A lance corporal (one stripe) was billeted nearby. He was about 6'8" tall and one of the most impressive men Mike had ever met. Cpl. North was conducting P.T. one afternoon and called out to Mike. "Hey! Yank! Get with it." From then on Mike was known as Yank. The barracks were probably 200 years old and were cold, damp, and not a good place to reside. There were two potbelly stoves in each bay, but Mike never saw them used. Probably due to a shortage of coal. The mess hall was not an impressive eating place. Meals still consisted of Bully Beef, potatoes, and oatmeal. Even though they were permanent billets they still ate from the mess kits. Mike was always a little hungry. Each evening a little old lady came through the area with some cookies for sale. Almost every night Mike spent two pence for a large cookie and never learned how the little old lady got the extra sugar to bake the cookies.

There was a pub Mike liked to visit. It was called the Boar's Rump. It was managed by a little widow woman. Other little old ladies would visit the pub in the afternoons. Many of them brought their pet dogs with them. The dogs usually drank beer out of ash trays which the manager provided. The pub owner was fond of Mike and usually saved him a pickled egg. Mike got along well with the other old ladies. Sometimes the pub owner would ask him to walk an old lady and her dog to her place of abode. They were very reserved and you seldom saw a drunk.

Mike was slowly developing friendships with the Canadians. One of the sergeants named John West, was especially prominent in the transformation of Mike from civilian to soldier in the Queen's Army.

It turned out that Aldershot was probably the largest garrison town in the world. With a parade ground over 300 yards long. All of the Commonwealth troops who went to England were there at one time or another. When one makes an ocean crossing with British Forces one automatically gets a seven day travel warrant, round-trip to any place in the United Kingdom. Mike selected Glasgow, Scotland as his target for a furlough. His first two tries were unsuccessful. His third was right on the money. His furlough had been granted. Upon arrival at Glasgow, Mike arranged for a bed through one of the service clubs. One clean cot for a quarter is what it amounted to. Mike purchased a round trip ticket for a cruise on Loch Lomond. The trip included travel to and from the loch on a double decker bus and a four hour ride on the loch. This crowd was in a fun-loving frame of mind, glad to be away from the danger of buzz bombs and other destructive means of waging war. They were buzzed several times by the RAF but it was all in fun. Several so-called buzz bombs had exploded in Scotland, doing little harm. A buzz bomb is fueled with a precise amount of gas. It is aimed at the target with just enough fuel to get there. In theory the engine cuts out over the target and causes it to be damaged or destroyed. A more sophisticated version homed in on radio broadcasts. This version was seldom used. Mike met some girls on the boat and obtained a couple of addresses. Mike was aware he attracted the attention of the girls and was glad that his shoes were shined and that he had altered his miserable battle dress.

Mike decided to leave Glasgow after four days and spent his last three days in London. He went to Charing Cross via subway from the rail station. He fell into a conversation with an AWOL American. Mike gave him two shillings for a bed that night and went on his way. The subways were full to the gills with people trying to find a safe place to lay their heads.

Mike's first concern was a bed. He located one at the St. Steven's Club near Westminster Abbey. His next concern was entertainment and he headed for Hyde Park. Mike spent the rest of the day listening to the soapbox orators expound on most any subject you could name. Mike especially enjoyed a debate between a little old lady and a callow youth with acne. They were fencing with each other about the sterling qualities of the British young people. The young man ended his tirade with the following "Dear Lady! I'll have you know that the British youth are the cream of the crop." The old lady responded immediately. "Let me tell you young man the cream of today is the cheese of tomorrow." The balance of his leave was spent in the central London area, Charing Cross, Westminster Abbey, and Buckingham Palace.

Mike finished his leave and returned to duty. He was exhilarated by his trip and wanted to talk about it with someone. The tall, Lance Cpl. North was nearby and had relatives in Glasgow and was ready to talk.

Cpl. North had taken his seven days in Scotland also. It so happened that on the way back he established a friendship with a war widow from Oxford and her four children. The corporal and the widow were near the same age and it turned out that there was a younger sister living with her. After two or three letters the widow invited them over to her place on a Saturday afternoon. Oxford is 30 or 40 miles from Aldershot and involved a short train ride to Reading and then on to Oxford by bus. It turned out that both women worked for the college. The younger woman had a rather important job in charge of a graduate program at the school.

They spent the afternoon just browsing around and purchased some fish and chips to take home for the evening meal. Fish and chips were plentiful and could be purchased without rationing. They were reasonably priced and tasted good. They were sold in an old newspaper wrapping. They were just a little greasy. About the only other food readily available was bread crumbs mixed with grease and made to look like sausage. So many potatoes were being consumed that they plowed up the golf courses and planted them in spuds.

The only meat other than horse meat that Mike ate while in England was rabbit. While he was in London, he patronized a black market restaurant that specialized in illegal food distribution. The rabbits were poached in Scotland. Mike was uneasy all the time but enjoyed the rabbit.

The widow had arranged for a baby-sitter and the four adults went to several of the local pubs. After two or three pints of dark ale, Mike and Margo began to thaw out and appreciate one another's assets. Margo was a well-developed miss and didn't hesitate to show her feelings and indicated her enthusiastic approval of Mike's presence. Mike's exposure to the opposite sex had been extremely limited due to strong supervision by his mother. Mike began to feel that he had missed something along the way and he rapidly began to make up for lost time.

The troopers' passes expired at 2200 and it was 2130 hours already. The time had slipped upon them. The last bus for Reading left at 2100 hours. There were no other busses until 0800 next morning. The boys left the ladies to their own resources and started hitchhiking. There was very little traffic at this time and no one stopped to give them a ride. A little after midnight a mail coach stopped and let them get on board. He could have been fired for this serious breach of security. It turned out that the mail coach was going through to Aldershot. The driver even poured them a cup of tea.

When they arrived at Aldershot they were about eight hours

absent without leave. This could be a serious violation considering the country was at war. The company commander was very lenient with them. He took the lance corporal's stripe and left him a private. He gave Mike extra duty for a week and restricted him to the barracks for ten days. The extra duty consisted of carrying coal to the kitchens and helping with cleaning the mess halls. Mike did not mind the extra duty, but the restriction was almost more than he could bear. He would not get to see his Margo for another ten days. Pte North had received no restriction and could come and go. He managed to get to Oxford and tell the sisters what had occurred.

The training accelerated. They spent several days on the range. The Lee Ensfield Rifle was pretty accurate. Mike knew he was a good shot, but didn't know how good. After firing several rounds from various positions, they took a break. A few minutes later a large corpulent sergeant major from brigade headquarters came puffing up. He asked to speak to Pte Casey. The sergeant in charge said, "You must be talking about the Yank." To which the sergeant major replied, "I don't care what you call him, get him over here." He had brought back Mike's target, and high and to the right of the bull's-eye was one of the neatest patterns you will ever see. You could almost cover it with your hand. The sergeant major told the sergeant to get the sights adjusted for both elevation and windage. From a prone position Mike fired several more rounds. The pattern was about the same, but almost all were in the black. The upshot of all this was that the innocent looking Yank was one of the best marksmen in the company and the New York state policeman was also in that select group. The same surprising results were all too evident when firing the Bren gun. The Bren gun was very versatile. It could be fired with a bipod from a prone position, with a tripod from a standing position, or from the hip. It was effective against low-flying aircraft. The Bren was a Czechoslovakian light, gas-operated field piece. It was made with the very best steel in existence at the time. It had a spare barrel which could be used at the drop of a hat. There was a barrel locking piece that permitted changing barrels almost instantly. A foot slogger could cool it with mud or piss mixed with dirt or any cool liquid. Heat did not appear to affect the weapon. Later they were permitted to wear the designation BG on their right sleeve. There was an opening for lance corporal due to the demotion of the big Cpl.

North. The New York state policeman was awarded the promotion and rightly so! He was a mature individual and handled himself well.

It had been rumored for days that the Queen and King were going to review the troops at Aldershot. Now the rumor was confirmed. The date was set for the following Saturday. This gave the troops five days for preparation. Shoes must be shined to perfection. Uniforms cleaned and mended.

Everyone was excited by the visit of royalty and all reacted in a different way. One of the Americans from Massachusetts must have found some loco weed. He was violent in the criticism of the event and very vocal with his objections. He said that the USA had a president whom every one respected and that the world would be better off if the royal family was removed from the world scene. He stated that he had every intention of going AWOL on the day of the visit. The rest of the Americans were pleased with the prospect of the visit and worked hard preparing their equipment and clothing for the royal presence. Saturday finally arrived and the entire company turned out with the exception of the unhappy American. He was listed as AWOL.

The review by the royal family was great. The number of troops involved exceeded 5,000. King George VI and Queen Elizabeth were very gracious. Queen Elizabeth noticed the USA patch on Mike's shoulder and shook hands. Then she asked what state he was from. He felt a mischievous streak and told her, "I am from Texas, ma'am." Mike never heard of the American who went AWOL. They know he got as far away as London. He jumped a subway gate and ran into the crowd. The next day, which was Sunday, found Lance Cpl. North and Mike in Oxford. Yes, Cpl. North had gotten his stripes back because of the simple reason that they needed him. The Sunday was spent just looking at the old buildings and other places of interest in Oxford. The town of Oxford was never bombed during the war. Two factors probably entered this picture. First many Germans involved in the war effort were educated in Oxford and second there were no significant military targets in the area. The university town of Nuremberg in Germany was never bombed either and probably for similar reasons.

The British were dead serious about their blackout blinds, even in a town that was never bombed. They carefully checked these every day. The block wardens were very serious about enforcement of the requirement. Motor vehicles were not permitted to operate at night, without special lights.

After the blackout blinds were closed for the evening, there was nothing much to do except visit a pub. There were good crowds every night, especially in the better establishments. The pubs closed about ten in the evening and the four pubhoppers returned to the girls' home. The house had a small rear stoop which could be called a back porch. Mike and Margo laid claim to this area. It afforded some privacy and was completely blacked out. Mike continued his education involving the opposite sex. They had arranged for a taxi to be at the bus station to transport themselves and four other soldiers back to Aldershot.

Training was proceeding at a rapid rate. Twenty mile marches with 30 lb. packs were commonplace. Night convoys of equipment and medical supplies were practiced. At least two Bren gun carriers were present in each convoy.

They learned to use the stirrup pumps as a means to control small fires set by incendiary bombs. Water would not extinguish a fire bomb. They were kept under control with sand bags and by spraying the surrounding area with water from the stirrup pumps until the incendiaries burned themselves out.

Mike enjoyed the drill sessions and learned to love the sound of bagpipes while marching on the parade ground. While in the orderly room to take care of some pay matters, Mike happened to notice a piece of correspondence with his father's letterhead on it. Knowing something was in the wind Mike said nothing. Mike knew his father well enough to know he was up to something. Mike, Sr. had in fact corresponded with the governor and former senator, M. M. Neely. Governor Neely contacted Cordell Hull, the Secretary of State. Mike, Sr. had started the ball rolling to get his son back where it was safe. He had followed the news on the Battle of Britain and was very much concerned. The wheels were turning in Washington, Ottawa, and Aldershot. The very next day he received a summons from Capt. Hunt and was questioned about the treatment he had received by the RCA. Mike told him that he had no complaints. The next day just after the troops had ceased their morning drill on the parade ground, two MPs accosted Mike and told him to be in the front of the orderly room in a half hour with all the clothing and equipment he had been issued. Mike asked about lunch and was told that it would be taken care of. The MPs took Mike to a large, ancient, well-kept building in another part of town. His Ensfield was taken from him and locked in a small closet. The rest of his gear was thrown into a corner of the room and he was sent to a mess hall for food. The meal was better than what he was used to, but it still contained the indestructible, indigestible horse meat. The dining facility was very quiet, most of the men appeared to have been wounded or impaired in some way, many were mental cases. Mike now knew why they locked up his rifle. There were several hundred disabled Canadians, many had been at Dunkirk. Mike spent an uneasy night at the facility.

The next morning he learned he was being shipped back to Canada with a group of war wearies. Space available back to Canada was extremely limited and some of the personnel had been waiting for weeks to get transportation home. One group with a high priority was the young men going to Canada or the USA for flight training. After two days Mike was sent to a facility about 30 miles from Liverpool. The camp was in a rural setting and very peaceful. A short walk through the woods brought one to a friendly pub where the troops drank ale and threw darts with the locals. No bombs or screeching sirens. It was wonderful.

Mike, knowing he wasn't going anywhere for at least a week, applied for a three day pass. Together with a weekend this made five days. Since the MPs called for him at Aldershot, Mike had not seen nor heard from Margo. He decided to use part of his leave at Oxford. Upon arrival he learned that Margo had promised to visit her folks near London for the weekend. She asked Mike to go with her. She wanted to get them acquainted with Mike. They traveled to the big city by bus and used the subway and rail system to reach their destination. Upon arrival at the station, near where Margo's parents lived, the air raid sirens began to howl, and Mike and Margo decided to seek the safety of a nearby bomb shelter. The station was practically deserted at this time of night and Mike and Margo were alone in the pitch dark shelter. The sirens were still screeching and Mike moved closer to Margo and attempted to embrace and comfort her. Margo gently demurred and pushed him away. They decided it would be better to stay in the shelter than walk to her parents' home. In the event the area was bombed, emergency traffic would make the streets unsafe. The two shadowy figures moved together in a tight embrace and this time Margo did not push

Mike away. About 30 minutes later the all clear sounded and a subdued pair of young lovers left the shelter and walked to the home of Margo's parents. The parents were nice folks and did not mind letting a homesick Yankee use their extra room. Mike returned to Oxford for the rest of his leave and then took the lonely ride to Liverpool. Mike stayed at the camp for another three days and lo and behold orders came for him to board ship for Canada.

It was rumored that the ship they went home in had just been to New Zealand. Regardless, the ship was well-stocked with such items as eggs and butter, ham and beef. About 100 RAF trainees were on the vessel heading for various bases in the USA for cadet training. The convoy was well-protected by destroyer escorts and a relaxed atmosphere prevailed. Eating, sleeping, and playing cards was the order of the day. Mike won consistently at cards. He enjoyed the company of the British cadets and ate like a pig. He gained six pounds on the five day voyage.

Upon debarking at Halifax, the troops were treated like royalty. Newspaper interviews, photographs—it was almost overpowering. Mike had developed a friendship with a Canadian soldier on board and was invited to stay at his home during the outprocessing. The trooper was married and had several children, one in the RCAF. A very homey atmosphere prevailed, making the transition back to civilian life more pleasant. Mike had learned that he could compete and survive in a hostile environment without sacrificing his principles. He had participated in important, historical events that helped shape the future of the world.

After two days of processing, Mike was given a furlough to return to Huntington for a few days. Mike wore his battle dress uniform and his "go-to-hell" cap with the black ribbon on it. He attracted a lot of attention while traveling and at New York he had a three or four hour layover. He was interviewed by reporters and radio people. All in all it was an exciting trip back home. Most of the family was at the railroad station to welcome Mike home.

Casey at about age two, ain't he a pippin'

CHAPTER 2

The Early Days

The earliest thing Mike could remember was seeing a work shoe covered with blood. The shoe was in a garage that housed an old Model "T" Ford. He recognized the shoe as one his father wore to work each morning. A few minutes later his Uncle Clyde came around the corner of the garage and covered the shoe with an old piece of canvas. Mike's childish curiosity was aroused and he asked what had happened. Uncle Clyde told him his father had hurt his foot at work and that Mike's mother was with him at the doctor's office.

The fact of the matter was Mike, Sr. was in the C & G Railroad Hospital with a foot so severely mashed it could not be salvaged and his young wife was in a serious, heated argument with the doctors about amputation. The doctors finally convinced the young couple that the foot should be amputated immediately. In fact, they had waited too long already. The foot and leg was severed about three inches above the ankle. The delay in amputation caused some serious problems. For one thing, gangrene had set in and another operation was necessary. In all Mike, Sr. had three operations during the next year.

A friend of Mike, Sr. who was a witness told how it happened. Mike, Sr. was operating a large steam crane. He had moved several lifts of engine wheels which weighed thousands of pounds. The lifts were critical and Mike wanted to make sure that things were all right. He dismounted from the crane to inspect the riggings. The hook must have been faulty. As he inspected the crane hook, it just straightened out and the wheels crushed him between the load and the loading platform, resulting in severe damage to the leg.

Mike, Jr. was seven years old when disaster struck and he grew up in a hurry. His father had always been a proud and extremely active man. Mike, Jr. could remember just a few months before when his father managed to carry his wife and two boys through the deep snow from the house to the old garage. Now he could not walk without crutches and it changed his personality entirely. Without the encouragement and loyal support from his wife, Mike, Sr. would not have been able to rededicate his life and change his career goals. After the stump of his leg healed he was in a deep depression. He could not function in his old job and with only a ninth grade education things looked dismal indeed. Sadie encouraged and cajoled him to go back to school. He entered business college and managed to graduate. The C & G Railroad Company gave him an office job and was happy with his work.

When Mike, Jr. was seven years old, he took his place as the man of the house. Dogs were killing the chickens and stealing eggs. This affected the cash flow and had to be stopped. Mike, Jr. killed three egg-sucking dogs the first summer after his father was incapacitated. Fences had to be mended and the garden had to be tended. Sadie had to spend all her spare time either at the hospital or tending to Mike, Sr. at home. Little brother William was too small (three years old) to be of assistance.

As Mike, Sr. began to recover he decided he would go into the chicken and egg business. He borrowed money from the local grocer and had two large chicken houses built and equipped for hatching and raising chickens. Chickens are very delicate and must have constant care and attention. Mike lost all his chickens and was back to square one. Although the chickens were gone, the chicken houses survived and came in very handy at a later time.

Mike, Jr. grew like a sapling. He started school several months early at Salt Rock while staying with his grandmother Ella. The first day at school he suffered a very humiliating experience. The school was equipped with outside toilets. Mike, Jr. had to go! He thought he could make it until recess and not have to humiliate himself by asking the teacher to go to the toilet. But this was not to be. He felt something warm on the inside of his legs and he let it all go. Grandma's house was near the school and Mike went home for a change of pants and a good ribbing by his Uncle Jim.

Grandma always bought flour by the barrel and one was now in evidence beside the kitchen cabinet. The household was relaxed. Some folks were napping and some were on the front porch talking. Mike was bored and happened to notice the cat curled up by the kitchen stove. Mike decided to find a better place for the cat to rest. He grabbed the cat with one hand and the flour barrel lid with the other. He threw the cat into the flour barrel and slammed the lid back on. The barrel was about half full and the cat had plenty of room in which to maneuver and maneuver he did. He let out a screech that could be heard in the next county and desperately began maneuvers designed to remove himself from the flour barrel. After some time the lid was dislodged and a cat who had drastically changed complexion leaped out screeching. The cat ran through the house, spreading flour all over. The entire household spent over one hour cleaning up the mess.

On another occasion, Mike incurred the wrath of his Uncle Jimmy. He had just sharpened and honed an ax until it would shave a grub worm. Mike was looking for something to do and the well sharpened ax caught his eye. He had been trying for weeks to get the metal wheel off of a wheelbarrow. Now was his chance! He confiscated the ax and started to work. After several mighty strokes he made an old ax blade out of a new one.

Mike enjoyed spending time at his grandmother's and went there every summer he could remember. He rode the old horse bareback, caught crawdads in the creek, and explored the farm. Mike's grandfather was the scholarly type and loved to read and didn't like to work. He had a trunk full of books and other publications which he kept at the foot of his bed. He was an itinerant preacher and roamed the countryside on foot preaching the gospel.

One summer Mike decided to play soldier. He made a fairly good imitation of a rifle out of an old broom handle. And in order to protect himself from artillery barrages, he dug a hole in the ground about three feet deep. The term "foxhole" was not a part of his vocabulary at that time. Mike just called it his private trench. He spent many hours playing war with his gun and trench. One weekend he went home. When he got back he noticed that he no longer had a trench. It had been filled up and a little mound of dirt was all that was left. He learned that a car had killed one of the dogs and his grandpa had buried the dead canine in the trench to keep from digging a grave. Mike took the broomstick which he had used as a rifle and turned it into a flag pole and tied some pieces of cloth to it and drove it into the ground. Grandpa passed the grave several times, but never mentioned it.

Mike's Aunt "Ellen" was about seven years older than he and they grew up together. Smoking was not considered a menace to one's health at that time. Mike decided he could not mature properly without a smoke. He mentioned this to his Aunt Ellen. She immediately set about procuring items she felt were necessary. Corn silk, paper, and matches were included. Ellen must have had some experience along this line. She cut the papers in little squares, cut the corn silks into smaller pieces. She rolled the corn silks into the paper. She handed the "cigarette" to Mike and struck a match. Mike accidentally inhaled the rancid smoke and started coughing his innards out. Ellen denied responsibility and said Mike did it all. Many years later she admitted putting the onus on Mike's back on several different occasions.

Every Autumn all activity ceased for several days and Grandma made apple butter. They had a large copper kettle and they would fill it full of quartered apples that had been peeled the day before. They would cook them all day. Someone had to stir the kettle all the time. Mike helped peel the apples. Aunt Ellen was helping him. One of the arms broke off the peeler while Ellen was peeling and she said Mike did it. Apple butter day was a long tiring day, but it was fun. There was quite a bit of tasting going on.

Grape vines were plentiful and grew near the roads. Mike and his buddies would cut the vines and make swings out of them. They would swing from a large tree and go out over the road. It would really startle the motorists. Some of them would get angry and stop. Others would keep going.

Another trick the boys played was to get a flour sack and fill it with leaves and tie a string to it and put it out on the edge of the road. Almost always a car would stop and someone would get out to pick up the flour. When they bent over the boys would jerk the sack and yell like crazy. It was great fun. A pocketbook would be just as effective as a sack of flour.

Mike did well in school when compared with the rest of his class. In a test conducted on sixth graders by the State of West Virginia, he was tops in his class with a 76, but top grades at other schools reached the high 80s.

One beautiful summer day, a catastrophic event occurred which had an influence on Mike, Jr. for the rest of his life. His parents had purchased an old, poorly constructed home adjacent to an industrial school for orphan black girls. One mile away was a orphan home for black youths. On this particular Sunday in

June the family was attending services at the Crossroads Methodist Church. They walked the one and a half miles from church to home. They were about half way home when they noticed thick, black smoke coming from an area that had to be near their abode. When they reached the last curve about 200 yards from the scene they could see the house almost fully consumed by the flames. Sadie was very distraught and upset. Mike, Jr. was sitting on the bank next to the mailbox crying his heart out. He had lost all his precious books and the erector set that one of the surgeons had given to him the last time his father was operated on. But his main concern was about his family. Mike, Sr. was still being rehabilitated after several operations on his leg. They had no clothing or any place to lay their heads that night. His brother, usually referred to by his mother as "little Billy" was too young to comprehend the full extent of the tragic occurrence.

The neighbors had managed to wet down the chicken houses with water from the well and creek. As a result the chicken houses were saved. Old Jack the family dog, who could whip any dog in the country, came to Mike, Jr. and started to lick the tears from his face. Very soon there was nothing left of the house. Sadie decided that the family should stay together as a unit and live in the biggest chicken house. There was much to be done and Sadie and Mike, Jr. started doing it. No one had ever seen Sadie work on Sunday, but they now saw her go to work with a vengeance. Mike, Jr. and some neighbors scrubbed the chicken house down and got rid of the acrid odor of chicken manure. Others helped wash the windows. The black lady in charge of the home for girls had some extra cots and bed clothes which she let the needy family use. People from the church brought in food and other items. Friends from the railroad company managed to get an electric line into the chicken house and "lo there was light." That night and for many nights to come the smelly chicken house was their home. Throughout the time (four months) they lived in the chicken house. Sadie gave thanks every night that she had a home and prayed that she would not have a problem with insects and nocturnal animals. After three or four days it began to look kind of homey. The wire cages where the eggs were laid was used as storage space and shelves. The chickens they had left in the other chicken house didn't seem to object to their new neighbors. They had a few problems with insects and other pests. All in all she managed to get along very well.

One of Mike's chores was the trip to the orphans' home over on the next hill. It happened about once a year. Mike noticed the cow was restless and mooing as if she were hurt. Mike, Sr. had already made arrangements for the visit and gave Mike, Jr. a one dollar bill to pay the farm superintendent for the service. The cow was to remain at the barn and Mike was to pick her up later. That afternoon Mike, Jr. put a halter around the cow's neck and led her home. He noticed that the cow was quiet and docile all the way back. Sadie never discussed the matter with Mike, Jr. but his pal Jimmie told him all about it.

Times were rough, the depression was just getting started. Mike, Sr. knew his family would not be able to endure the miserable winter months in the chicken house. He borrowed money from a local grocer, who was a friend of the family. Men from the C & G Railroad helped clear a building site and fill in a part basement left when the old house burned. There were no house plans, only rough sketches to build a sturdy two bedroom house. The carpenters were capable and eager to help the family get a home. They pretty much planned as they went, and had frequent conferences with Sadie. About the first of November the house was ready for occupancy. A significant event which occurred about this time was the stock market crash of 1929. It was a long, lean winter. Payments on the loan had to be made. Mike, Jr. had picked a lot of blackberries and Sadie had canned them along with several gallons of apples. They had killed two hogs and salted down several bacon slabs. Mike, Sr. had tried to cure the hams with only partial success. He was very much disappointed when one of the hams went bad. Mike, Jr. milked the family cow twice a day. So! The family had ample milk. The family, while lacking in some ways, had abundance in other ways. They lived on cornbread and milk, pinto beans, oatmeal, and pork. Both boys were healthy and active. Mike, Jr. walked about one and one half miles to school and back. He carried water from the well to the house, took care of his school assignments, and did other chores as required.

Saturdays were always a big day in the lives of the family. Mike's Uncle Joe usually came out on Saturdays and Sundays, but Saturday was special in more ways than one. Uncle Joe was a salesman for Brach's Candy Co. and he always had plenty of samples. Saturday night was grocery shopping night. The family would go to 20th St. to purchase groceries for the coming week. Many staples were bought in bulk such as rice, beans, oatmeal, and ground meal. Trips to the grocery stores were also part of their recreation. Occasionally they would purchase beef brains and liver. The brains were considered a delicacy when fried with eggs and they were not prohibitive pricewise.

Some of the older boys in the area had developed a game they called "grab". When a newcomer moved in they always talked him into a game of grab. The game was explained to him in such attractive terms he just couldn't resist. All of the boys liked good pocket knives and most of them had one. They generally took the newcomer to the pasture fields back of the school and pulled out pocket knives. They explained that each boy who wanted to play needed a pocket knife to put up. They put all the knives together in a pile and blindfolded the contestants. Usually about five or six boys would play. They gave the newcomer a chance to go home and get a knife if he didn't have one. They left the blindfold on the victim and the other boys took theirs off. Normally the victim picked out an old, cheap knife to play with. The victim could see the knives all in a pile before he was blindfolded and some of them were very good. They put a bushel basket over a cow pile. Everyone knelt on the ground around the basket and someone took the basket off and yelled grab. Greed got the best of the victim. He couldn't resist getting something for nothing and he was usually fed up with his old knife. The victim was the only one who grabbed and he got both hands full of soft cow dung and still had his old pocket knife.

About this time Sadie determined that her boys needed a trip to the zoo. There was a good one at Cincinnati about 175 miles southwest of Huntington. The C & G Railroad Company provided passes for their workers' families and if one packed some food, travel was not expensive. Sadie got everything ready and they left one morning about 4:00 a.m. for the four hour rail journey. Upon arrival at Cincinnati about 8:00 a.m., they caught a streetcar to the zoo. The boys were aware of most of the animals from seeing their pictures in the National Geographic subscribed to by their Aunt Gertrude. The zoo closed at 5:00 p.m. The train left for Huntington at 9:00 p.m. So! They had to spend several hours at the station. Sadie didn't want the boys out of her sight and was always scampering after them, especially "poor little Billy". They took these trips for the next several years.

One interesting event took place about this time. Mike was riding a horse bareback through the yard at Salt Rock. There was a clothes line strung along the yard and Mike had forgotten about it. He hit the clothes line about neck high and it knocked him off the horse. That wasn't the first time he had stuck his neck out and it wouldn't be the last.

Mike, Sr. was making remarkable progress in his adjustment from a normal person to an amputee. There were many examples around the community where tragedy had struck and the victims had given up any attempts at rehabilitation. Mike, Sr. tried all the harder. He was constantly trying to improve his lot and set a good example for his two boys. About 1932 he helped form a Boy Scout Troop and served as the first scoutmaster. The boys worked and made money to buy uniforms. They formed Troop 24. They met in the cafeteria of the school. There were 13 boys in the troop when it was formed and it grew to about 18 boys within a year. At a camporee in 1933 Mike, Jr. won first place in a contest to determine the champion in building a fire without matches. He built a fire with flint and steel and had half a gallon of water boiling in six minutes.

Mike, Jr. also entered a soup making contest. He figured that if he could boil water he could boil soup. The first round of soup making eliminated all but Mike, Jr. and a scout named Tom. It was decided that the troop would eat soup the next day for lunch and that the two contestants would provide the meal, consisting of soup. And the one cooking the best soup would be the champ.

The next day both contestants were busy making preparations. Tom's father chewed tobacco and Mike, Jr. knew where he kept it in his truck. He appropriated a generous portion of the tobacco to help flavor Tom's soup. While Tom was busy peeling potatoes, Mike, Jr. dropped all of the tobacco into Tom's soup. The stew just bubbled a little bit and changed color slightly, but for all practical purposes it was a good mixture. The two scouts declared that dinner was ready and to let the judging begin. Three of the parents were selected as judges of the stews.

Mike, Sr. excluded himself from the voting for obvious reasons, but said he would vote in case of a tie. The judges

tasted Mike Jr.'s soup first and found it to be very good. They tasted Tom's soup and appeared to be bewildered. One of the judges said it tasted like "Gar Broth" a term Mike's mother used often on something she didn't like. The other judge owned the tobacco and thought it tasted fine. Since they could not agree Mike, Sr.'s vote was called for. He took one look at Mike Jr.'s face and declared Tom the winner. He knew Mike Jr. had compromised the soup in some way. Mike, Jr. admitted his role as the spoiler in the great cook-out. Needless to say, there was not enough of the untainted stew left to feed all the scouts. So! Several cans of pork and beans were opened and consumed.

During the Summer of 1937 Mike, Jr. worked with the C & G Railroad Company in their lumberyard, he made 40 cents per hour and earned every penny. The company stacked the lumber and stored it until it was needed. Then they shipped it all over the system. They devised a way to handle the lumber by balancing it on a greased pole and sliding it into place. It was hot, dirty work, especially inside the cars. People of all ages worked on the crew. At lunch time they generally separated into three groups. The older men sat at the tables near the office and played set-back (card game) and talked about what a good bowel movement they had that morning. The middle-aged men generally discussed their wives' cooking and what a good meal they had the night before. The young men got together and sprawled out on a stack of lumber and told about how many women they had seduced the night before and all the grisly details. They usually took a short nap.

One of the men who worked on the crew was huge. He weighed over 350 pounds. One afternoon he was inside a railroad car unloading 2 x 6s and got hot and dirty. He took off his bib overalls and was working in his shorts. Mike picked up his overalls and ran down the tracks. The big fellow ran after him, but he didn't stand a chance. Mike ran to a signal post about 15 feet high and hung the overalls about ten feet up. The big man was fit to be tied. He could not climb the pole because of his weight. He just stood there and cussed. Finally he got one of the younger men to get them down. The young man fell during the climb. Interestingly he was not hurt, but Mike, Jr. received his first lesson in safety. Horseplay on the job should not be condoned. Mike, Sr. was always teaching Mike, Jr. how to use dangerous tools such as axes, crowbars, shotguns, knives. Mike grew up with respect for the safety of others and developed an interest that would someday earn him a good livelihood.

Mike, Jr. was 15 before he was permitted to attend a party where they played kissing games. At every party he attended Sadie was right there. They set up a small closet as the post office and the game started by the host or hostess sending someone a letter and when they came to get it they got a big kiss instead. Mike, Jr. waited anxiously for his name to be called, but Sadie's presence more or less prevented this from happening. Finally one of the more liberal girls wanted to see what Mike, Jr. was like close-up. She called his name. Sadie said, "Mike, you don't want those slimikie girls slobbering all over you now, do you?" Mike looking a little sheepish went into the post office. After about 15 seconds Sadie said, "That's enough of this Mike! Come on out of there." Mike came out looking as if he enjoyed being slobbered on.

Mike, Jr.'s Sunday school class had a picnic at Lake Vesuvius in Ohio. There were about 20 young people in the class. All concerned had food items and were in a holiday mood. There was a large parking lot near the picnic ground. The picnic was just getting started when around the hill a horse appeared. There was no one on it and it seemed to be trying to shake off its saddle. Two boys from the nearby stables came over and caught the horse. They said his name was Bucky and that they had just rented him to another young man who could be seen limping around the hillside. The girls were listening to the tale of the bucking horse as if fascinated by it. Mike, Jr. said that he would like to rent Bucky. He wanted to show the girls what he could do. The boys said that if Mike, Jr. could ride him there would be no charge. Mike climbed on Bucky and he took off up the row of cars running real close to them in order to bump Mike, Jr. off. Someone with a model "A" Ford coupe had left the rear of the car sticking out a little farther than the others. The horse hit the rear end of the Ford which caved in the rumble seat. The impact of the horse's kick jolted the Ford into the next vehicle, a Plymouth, and caved in the Plymouth's rear door. Mike, Jr. was thrown over the top of the Plymouth and landed on another Ford broadside and put a dent in it. There he was with three wrecked cars and a bucking horse. The story was stretched every time it was told and became legendary around Beverly Hills.

During the '37 Flood, Mike spent three days away from

home on flood duty. He was very effective transferring people from their flooded property. He went to the West End of Huntington and volunteered his services. When they asked for a river man to help evacuate a dwelling, Mike, Jr. proudly stepped forward. He had been on the river several times and was a good swimmer and was not afraid. He was given a boat and a set of oars. He was told to go help evacuate a family. He worked till dark and slept in an abandoned stove factory. The Salvation Army fed the volunteers with sandwiches and coffee. Mike, Jr. felt sorry for the flood victims, but he enjoyed the activity. The last day he worked was with a minister. Mike, Jr. took him every where he wanted to go. The minister was a very energetic man and kept Mike, Jr. busy. The clean-up took several weeks and resulted in a floodwall being built to protect the city.

When Mike, Jr. graduated from Huntington High School with a "C" average, he had little to do and nothing to do it with. President Roosevelt had set about creating relief agencies like the WPA. (Works Progress Administration), CCC (Civilian Conservation Corps), NRA (National Recovery Administration), and several more. Mike, Jr. decided to join the CCC. In the meantime, Mike, Sr. had announced for the West Virginia Legislature for a two year term and his prospects for getting elected seemed good.

Mike, Jr. went on his last camping trip with the Boy Scout troop. They were near Salt Rock for about one week. They erected tents and had a cozy campground right on the river. The camp was well-organized. The scouts made their own rules and regulations. There was a new sheriff appointed each day. The sheriff had powers of arrest, but for most offenses he used the sheriffs book and wrote about the violation. The scoutmaster appointed the first sheriff. And as each sheriff finished his tour, he placed the sheriffs book under the pillow of a scout and no one knew who he was designating sheriff. This activity went on for a full week and worked well. The sheriff, took the job seriously. At the end of the day court was held. A judge was appointed and also a prosecuting attorney. A scout master or commissioner monitored the activity to keep it on the right track. The entire troop was the jury. The most common violations were throwing trash around the area, stepping in the tent ditches, and getting in the river when no adult was around. Sometimes these were circumstances that called for a not guilty. But most of the charges led to convictions. Sentences were passed and enforced on the guilty. Typical punishment included: KP duty washing the pots and pans, picking mulberries which were plentiful, carrying potable water from a nearby farm house, gathering wood for the evening campfire, etc. Mike, Jr. was on KP one afternoon and he let some dogs lick all the food off the dirty plates. The sheriff caught him. At the trial Mike, Jr. stated that he intended to wash the dishes in soap and water later and they would be clean. The jury convicted him and he had to serve an extra tour on KP duty.

Mike, Jr. went to the agency that recruited for the CCC and learned that his father had a good friend in charge of the office. Mike, Jr. filled out all the necessary papers. When he left, the official in charge of the program called Mike, Sr. and asked how he should handle the situation. Mike, Sr. requested the enlistment process continue, but to keep Mike, Jr. at a location near home. For the second time politics had reared its head in his life. Mike, Jr. was sworn in with about 100 other recruits. Ninety-nine of the 100 went to Fort Knox for transfers West and Mike, Jr. was sent to Camp Kanawha in Charleston. Mike, Jr. was fit to be tied. One of his main reasons for enlisting was the probability of travel to the West. He was more determined than ever to get an assignment as far away as possible. Mike, Jr. went to Charleston by train and was met at the station by a 6 x 6 truck about 9:30 p.m. Upon arrival at Camp Kanawha, he was assigned a barracks and issued two sheets, a pillow case, and a blanket. The lights were out in the barracks when he arrived. There were two empty cots with the mattresses folded. Mike, Jr. selected one of these and made his bed in the light of the CQ's flashlight. Full of misgivings and more than a little homesick, Mike, Jr. fell into a troubled sleep.

The barracks held about 20 people and most of them were gathered around the area where all the excitement was occurring. Wham!! Bam!! Mike, Jr. woke up quickly. Someone had turned his bunk over with a vengeance. One of the men was straddling his bunk waving an old beat up automatic pistol and yelling like crazy. It was a disquieting scene. Mike, Jr. began to straighten his cot up and a 220 lb. slob with a big gut turned it over again. This time Mike, Jr. was in a different position—instead of eight or ten screaming hecklers, he had just one tormentor to concentrate on. And concentrate he did. He was fighting for existence and the right to carry on legitimate activities in any way he saw fit. He hit "Pus Gut" in an area around the kidneys and stood with his guard up awaiting another attack. The barracks chief noting the disparity in size and age was afraid someone would get hurt. So! He declared a truce until that evening when some disputes were settled by the use of boxing gloves. Mike, Jr. spent the day getting clothing issued, getting shots, and getting familiar with the camp. He was assigned to a different barracks with more bunk space. The fight was to take place in front of the mess hall after everyone had showered and had gotten ready for chow. Mike, Jr. weighed 170 lbs. and was severely out gunned. The First Sergeant hit on a dishpan to start the melee, which was to last three two-minute rounds.

Mike, Jr. moved easily around his cumbersome opponent getting a few jabs and dancing away. His adversary was making wild swings, trying to end it with one blow. He finally hit Mike, Jr. in the neck, knocking him down. As the fight went on "Pus Gut" began to tire and Mike, Jr. got in some good punches. The fight ended with both fighters saving face and Mike, Jr. was quite a celebrity with the newer and younger recruits at the camp.

The camp commander, a "Lt. Burk", learned that Mike, Jr. could type. He also noticed he was careful in his dress and made him the company clerk. Mike, Jr. opened the mail, typed letters, filed orders, and kept the office clean. For this he would get an extra $6.00 per month. Mike, Jr. didn't care much for the job, but did well.

The camp sent a recreation truck into Charleston each weekend. Mike did not go to Charleston and opted to spend his time at camp. He made short fielder on the softball team. There was a church about one mile down the creek. They usually held church two times a week and Mike attended quite often. He walked home with several of the girls and became acquainted with their families.

After about two months the camp, which was located in Wayne County, was scheduled to move West. The organization needed ten more men in order to be at full strength. Letters were sent out to all camps in the general area, asking for volunteers. Mike, Jr. was beside himself when he opened the mail. Needless to say he should not have read official letters without permission. Mike, Jr. planned his campaign and started to work right away.

Casey at the Indian reservation near Moapa, Nevada.
Just had his head shaved.
He never did that again.

He knew the first aid attendant who took care of the dispensary could type and was dissatisfied with his job. An ambitious assistant could fill the first aid job.

Mike, Jr. had a very good relationship with the First Sergeant and had solved some problems for him. The First Sergeant agreed to assist him in his endeavor to transfer to Cabwaylingo and move West. Things moved like clockwork. The camp commander agreed to include Mike, Jr. in the transfer orders. The very next day ten men from Camp Kanawha boarded a truck and set out for Cabwaylingo. Route 60, the direct route to Wayne County, went by Mike, Jr.'s home. The driver of the truck agreed to stop for Mike, Jr. to inform his mother of recent events and that he was headed West. Sadie was quite upset over the turn of events and wanted Mike, Jr. to call his father. Mike, Jr. tried to call but Mike, Sr. was not in his office. The weapons carrier proceeded to Cabwaylingo where two days later they boarded the train for Moapa, Nevada. Since Mike, Jr. had never been farther away from home than Cincinnati, Ohio, the trip was a revelation to him. He had a little box camera and took several pictures of the trip. The town of Moapa consisted of four buildings. One was a small jail about 10' x 10' used mostly for drunken Indians. There was a reservation nearby. Another was a typical western bar with slot machines and all the trimmings. The third small building housed the railroad station and the fourth building was a residence for the section hand on the railroad. The young men lined up for roll call and were then marched over the hill to the camp named "Muddy River." The men were assigned to their various barracks.

Many of the men who transferred west from Cabwaylingo were from the Huntington area. Mike, Jr. had attended school with some of them. They had several things in common, shortage of cash and not much training. All were glad to be in Nevada and were optimistic about the future. Mike had the urge to do something drastic. He talked the camp barber into shaving his head. The hair grew out in a hurry and Mike never had it shaved again.

There was a spike camp that housed 20 or 25 men from the main camp. This spike camp was located about 35 miles away in the virgin mountains. Mike, Jr. helped take a load of supplies out to the camp and it was love at first sight. He was determined to

The camp at Moapa. Casey lived here for one year and operated a bulldozer for the CCC.

get assigned to the camp. On the train ride out, he had become acquainted with Aaron Clay, a bulldozer operator. Mike, Jr. took advantage of this budding friendship to request a spot as a grease monkey. When 20 men were chosen for the mountains, Mike, Jr. was on the list. The camp had eight sleeping tents with four bunks in each, one dining tent, seating 30 people, and one cook tent. A cellar dug into the side of the mountain kept perishable foods cool. Local supervision was provided by a 50 year old project assistant who also served as cook. A project manager came every day to supervise the work which consisted mainly of building drift fences, building roads, and developing springs.

There was a young man about Mike, Jr.'s age and size at the spike camp. He was bitterly disappointed when Mike, Jr. got the dozer operator's job. He showed this dislike whenever he could. One day he left an oily rag on the seat. It was dirty oil and soaked into the seat, and Mike, Jr. called his hand on it. The young man lost his cool and called Mike, Jr. a name not complimentary to his mother. Mike, Jr. reacted in the only way he could. They fought like tigers in the 7,000 ft. altitude and quickly tired out. The other youth apologized to Mike, Jr. and they called it quits.

Equipment included an International Harvester T-40 diesel bulldozer, a manually-operated grader, a generator, and jack hammer. Mike, Jr. was determined to become the bulldozer operator when Clay's enlistment ran out. He learned about how to change angles on the blade, how to lubricate and change oil. He operated the dozer every chance he got. Clay had been an operator for three years and knew his stuff. Every morning the dozer crew removed the tarp and greased the equipment if need be. They then carefully started the small gasoline engine to provide heat to warm the big diesel so it would start.

Life at the camp was very different and exciting. Animal life was visible all over the area. Jack rabbits were very much in evidence. Birds somewhat like our West Virginia quail were plentiful. Small deer could be seen at times. Coyotes were all over. And Mike, Jr. became used to their mournful howl at night. Gila monsters (large, poisonous reptiles) dozed in the sun. There were scorpions galore with their poisonous tails arched over their backs. Snakes were always a problem. The area was full of rattlers. Bull snakes abounded. Mike, Jr. always had a bull snake around as kind of a pet. Bull snakes look somewhat like

Casey with his bull dozer crew. He is on the far left, no shirt.

rattle snakes, but were not poisonous. Rattlesnakes were most common and most feared, rightly so! They could be deadly. The project assistant and cook, called Shorty, kept several rattlesnakes at a time. He made belts and hat bands out of their skins and sold them. Shorty had an orange crate covered with a fine mesh wire and kept three or four snakes in the cellar. Each time one of the men was sent to the cellar for some type of provision, the snakes would rattle their tails off. The cellar became a place where no one wanted to go. One afternoon when Shorty went to the main camp for supplies, three of the hardy souls, including Mike, Jr., decided to kill the snakes. They had a generator which ran most of the time and put out thick, rancid, black smoke. The snakes were carried out, and the box carefully placed under the exhaust fumes of the generator. After two hours in the fumes the snakes were not affected. They had more energy than before and were rattling up a storm.

Shorty would cook the rattler flesh in a skillet along with some other meat and serve it hot. It is a pretty meat, pinkish in color and no fat. The cook fried rattlesnake any time he needed a snake skin to make a hat band or a belt. He wouldn't waste a snakeskin just to eat. He sold a lot of rattles and snake skins to people in the Las Vegas area.

Several Indian tribes had occupied the general area in past years. Hieroglyphics were common on the cliffs. Old habitation could be found in caves and under cliffs. Smoke stains from old cooking fires could still be seen on the large rock formations. Mike, Jr. began collecting Indian artifacts and pieces of quartz along with pottery shards from old campfires. The caves where Indians had lived were easy to discover because of the peculiar composition of the dirt floors. It was more like fine dust. The material lent itself well to a screen.

Mike, Jr. built a three foot square screen and sifted through many tons of material. On weekends when the others had gone to the main camp for the weekend, Mike, Jr. took his screen and shovel and plenty of water and went looking for artifacts. There was a female donkey (jenny) that hung around the camp. Many times Mike, Jr. loaded her up with his gear and hit the mountains. The powers that be sent a project manager to the camp. He was a bachelor about 60 years old, named Metcalf. He had some experience mining for gold. His father was rumored to have made quite a gold strike in the Music Mountains. The first

Jenny in the kitchen
Casey used her to haul his equipment in the desert.

weekend Metcalf was there he noticed Mike, Jr. heading out into the semi-desert country with the jenny. The old man asked Mike, Jr. if he would like to look over more country than the jenny could travel. Metcalf had been told to familiarize himself with the country around the spike camp and was permitted to use a government vehicle. Mike, Jr. switched his gear over to the weapons carrier and eagerly went with the old man. They traveled to Gold Buttes, a place Mike, Jr. had never visited because of the distance. The weapons carrier had shovels and other equipment aboard and could travel the rough desert country pretty well. The only building at Gold Buttes was an ancient clapboard dwelling. Two old men lived by themselves in the house. The place was completely surrounded by empty whiskey bottles. The men must have carried them in from other places as well as emptied many themselves. Mike, Jr. and Metcalf got well-acquainted with the old men and visited them often. They panned for gold at several locations near Gold Buttes and found some color. Locating water to pan with was a real problem. Occasionally they found some water and other times they carried it in milk cans. What little gold Mike, Jr. panned he put in a small glass about the size of a small fountain pen. It was a container used to hold a styptic pencil of the type one uses to treat small cuts from shaving. Mike, Jr. accumulated about a fourth inch of gold in the small glass. Later Metcalf quadrupled the amount of material processed by making a small metal rocker and using a chamois skin to catch the gold-bearing sands.

Mike, Jr. still followed his hobby of discovering Indian artifacts and one memorable afternoon sifted three perfect small Indian arrowheads through his screen. Great care had to be exercised to keep from getting snake bitten. The Gila monsters were too slow to be much of a threat, but the rattlers were very active. Mike, Jr. still has a fine collection of Indian arrowheads.

Many of the men felt isolated and bored at the spike camp, but young Mike, Jr. was thrilled by it. He inherited the job of operating the dozer. He arose each morning at 5:30 a.m. sharp. He and his helper had breakfast with Metcalf. If the dozer was parked nearby they walked to its location. When the dozer was left out in the field, Metcalf took the dozer crew to work in the truck. After Mike got the dozer started he stripped down to his white jockey shorts until all he had left was his socks, shoes, and shorts. He was brown as a berry. He did not have knotted muscles, but was very well built and jumped around over the rocks like a mountain goat. Sometimes he would get a mile away from the grader crew. The grease monkey would find a shady spot and Mike would operate the dozer solo. One morning he was cutting banks on both sides of a dry creek bed. On the bank was a hole obviously used by an animal. Mike was told not to veer too much in building the road, but to go according to the stakes. He got as far from the hole as possible and yet remained inside the stakes. The hole curved back to the left and was a coyote nest. Mike had no way of knowing this and dug up three young coyotes. They started sounding off and were not hurt. Mike went down the road about half a mile and started another segment. In the meantime the mother coyote hid her young in another location. Mike went back and finished up grading the creek bank and left it to the scraper crew to finish. At several locations in the area, university students and other folks were conducting digs for dinosaur bones. Mike and Metcalf had many pleasant visits with them at the various sites.

Barrel cacti were plentiful in the semi-desert country around the camp. Mike soon learned that in case of emergency the barrel cacti would serve as a reliable water substitute. Just cut a circle around the top of the cactus. Lift off the spiny section that makes up the tuft of the plant, pound it with a smooth rock, make a small indentation and liquid will gather in the resulting reservoir.

The spring was always beautiful in the desert. Small desert flowers by the millions made it a very rare and unusual place. In the 1930s people were few and far between in the Virgin Mountains. Recreation vehicles were almost non-existent. There was no trash or junk cars in the area. Delicate plant life had not been disturbed. It was pretty much that way until the early 50s. Nothing is more destructive than uncaring people. The wheel tracks of vehicles destroy delicate plant life. The desert was home to various Indian tribes and artifacts were often found in caves and hieroglyphics on canyon walls. Most of these sites have been destroyed. The cacti were deceiving. The most beautiful were often the most painful. If you touched them they left small needles in your skin and could cause severe infection. This was nature's way of protecting a delicate species.

The main camp operated a well drilling rig and drilled many wells for the benefit of the ranchers. There was an active drilling

Lunch time in the desert. Casey appears hungry.

program at the Rex Bell Ranch near the Joshua National Forest near Search Light, Nevada. Two men were assigned there to drill a string of water wells. One of the men was a contractor, the other was a CCC enrollee. They needed to get the CCC member back to the main camp for shot record check and discharge processing. They needed someone to take his place for about two weeks on the drilling rig. Mike was head of the dozer crew and believed that his number two man could handle everything for two weeks. He decided to volunteer for two weeks at Rex Bell's Ranch. Rex was a former movie star, and had an eighty thousand dollar home. That same home would probably cost 1.5 million at today's prices. He piped in water from a mountain nearby. The home obtained power from two generators, one a standby. There was a bunkhouse for the cowhands and that is where Mike lived. The bunkhouse had a kitchen and about six different bunkhouse rooms with a porch running all along the front. Mike noticed a tarp with a hind quarter of beef wrapped up in it. Mike asked one of the cowhands about it and was told it was there for all hands and if Mike wanted a steak to just slice off what he wanted. Mike ate a lot of prime beef while he was there. The CCC camp mess provided them with sugar and carnation milk and several cartons of small cereal packs. The cowhands ate the cereal and liked it. One of them ate it three times a day. The dry desert climate did not have a bad effect on the beef and it was always there in abundance. Much of the time the drill ran itself and there was nothing much to do. When they did work it was hard, bone-tiring labor. After two weeks Mike returned to the camp and continued working the dozer.

Several of the men who could afford it planned a trip to Pasadena, California, to see the Rose Bowl parade and game. It ended up with four going including Mike. They rode the train to Los Angeles and got rooms at a cheap Chinese hotel. They spent some of the time just walking the streets, eating in cheap restaurants. The next morning found them in Pasadena getting some space along the parade route. They viewed the Rose Parade and proceeded to the football field. The big game was played with Duke University and Southern Cal. Mike was lucky he wore a sweater with a big "H" on it. A man and his daughter were standing talking. The man saw the big red "H" on Mike's sweater and was told it was for participation in sports at Huntington High School. Mike, Jr. was a member of the track team. The man was a high school football coach and had purchased three tickets many months before. He had his daughter with him, but the other person did not show.

It was almost game time and the coach said he would let Mike, Jr. have the ticket for exactly what he paid for it. It was a real bargain. The man could have easily sold it for $20.00. The ticket cost Mike, Jr. $4.40. One of the CCC boys paid $18.00 for a ticket. The game was extra good.

Mike was glad to get back to his bulldozer. He spent the rest of the time building roads and continued his exploration for Indian artifacts. Sometimes on Sunday Mike and Metcalf, who was a Mormon, drove the truck to Overton, Nevada and attended the Mormon church there. Metcalf took Mike, Jr. to the homes of several of his friends and Mike, Jr. was pretty well acquainted with the area around Overton, Nevada.

Mike began to learn something about the game of poker. It was against the rules to play poker in the camp, but they did it anyway. Sometimes after lights out several of the men would gather in the latrine for some stud or draw poker. One of the players, a man in his thirties, was almost a professional. He knew when to hold them and when to fold them. He bluffed quite frequently, usually at the right time. Mike learned the game very well and it paid off over the years. One memorable night Mike drew two cards to 3-4-5 of Hearts and drew the two and six of Hearts making him a straight flush, almost unheard of in draw poker. It was a pot limit game which meant you could bet the size of the pot. The pot opened with one of the six players betting three dollars, the size of the ante. Mike used poor judgment calling and drawing two cards. He was losing and wanted to get even. The professional player drew a full house and bet $21.00, the size of the pot one other person called. Mike had $39.00 in his stack and he bet it all. The professional and one other man called. The pot totaled $142.00. Mike got all his money and then some. He learned not to play out of his class and most every month he would send some money home. The clean desert air and early to bed and early to rise living plus the good food available put Mike in mint condition as indicated by the pictures he had collected. In September Mike got a letter from his father telling him he could get a job if he wanted to come home. Mike had been sending $25.00 a month home and his parents had bought some property in town. Mike, Sr. was in the legislature and the future looked good for the family.

CHAPTER 3

Pre-war Years

Mike was sent to Fort Knox, KY for discharge and became a civilian again. He got a job with a contractor at South Charleston, WV, the place of his birth. The foreman on the job learned that Mike had handled a jackhammer and put him to work on it. Later that afternoon Mike dislocated his shoulder and went back to Huntington. When he came home, little Billy had grown up and was popular with the girls around home. Sadie followed him around like she had Mike, but Billy managed it better. Mike gave Billy a fountain pen and a silver dollar. At that time a dollar was a large amount of money. Two days later, one of his buddies had the pen and the dollar wasn't there anymore. The boys had a small poker game going out behind the church. It was really a small game, mostly pennies, nickels, and dimes. Most of the gamblers were about 16 to 18 years old. On occasion an adult would sit in on the game. The boys would argue around and play all day to win a few pennies.

One morning Billy and about four other boys were playing stud poker. The largest bets were never more than 10 cents. A stranger came through the woods and asked if he could play. He pulled out 10 or 12 dollars and started to bet 25 and 50 cents at a time. Everyone was glad to see the new blood in the game, especially since he had some new money. He said he was recently out of the penitentiary at Moundsville. In about an hour he had lost all his money and left the game. The game broke up a little while later and Mike went home. Sadie came in fuming. She told Mike that she had been out visiting the neighbors on the corner when the neighbor's nephew came in and told them he had just lost all his money to a card shark. About that time they saw Mike walking up the road and the nephew said, "There's the SOB that got all my money!" Sadie said, "I never was so embarrassed in all my life." It seems that when the nephew got out of the pen they gave him a bus ticket and $10.00. Mike had his $10.00 and felt a little remorse for embarrassing his mother, but did not give the money back.

Just prior to the Florida trip, Mike, Jr. had entered an intermural boxing contest. He was in pretty good shape physically and he decided to become a big time boxer. He entered a boxing match sponsored by the physical education section. He worked hard at trying to get in even better shape. He just barely made the weight requirements for the heavy weight division. His opponent weighed about 215 lbs. The tournament took place at the college gym which had a track around the outside of the gym. About twenty feet above the boxing ring, all the boys from the village were all lined up around the track there to see Mike get thumped. There were three two minute rounds for each fight. The other fighter belonged to a fraternity and they were out in force to root for their man. The fight was going well until the third round. The Beverly Hills boys were very vocal and urging Mike to give it his all. Some others even risked a quarter on him. Mike let go with a wild haymaker and his shoulder became dislocated, the same shoulder he had dislocated before. Fortunately, the referee was on the ball and stopped the fight before Mike could hurt himself. He was in such pain he couldn't move. They took him to the hospital and set the shoulder. Mike went home with a wrapped shoulder. Sadie heard him groaning during the night and went to check on him. She saw the bandage and became very upset. The shoulder gave Mike trouble for most of his life. When the arm got better, Mike decided not to delay the trip to Florida.

Mike, Sr., through his political connections, got an alternate's appointment to West Point for Mike, Jr. The appointment was made by then Senator M. M. Neeley. Mike, Sr. obtained a rail pass from C & G Railway for Mike, Jr. to travel to Washington, D.C. for the physical and mental exams. Senator M. M. Neeley was a rough and tumble politician from West Virginia. He served one term as governor of the state. Mike, Sr. and he became fast friends. Neeley appointed Mike, Sr. to the World's Fair Commission and he and Sadie spent some time in New York taking care of state business.

Mike, Jr. stayed in Washington three days and failed the written exams. The failure was due to a lack of basics such as math and geometry, algebra, and the fact that he did not have the advantage of a prep school. Mike Jr. never discussed the failure to pass the test with his friends. Mike enrolled at Marshall College majoring in journalism. He put in a semester with very poor grades. He quit college after the first semester and decided to go to Florida. This was June 1940. Bill was supposed to graduate in June from Huntington East High School. He turned in his cap and gown and got the rental fee back. He sold his books for a few dollars and was ready to go, but still did not have much money. Mike had about $90.00 left after he bought a model "A" convertible Ford coupe. He gave $30.00 for the car and some of his friends helped him tune up the vehicle. The model "A" was painted a fire wagon red and had a rumble seat.

Sadie didn't approve of the trip to Florida, but helped get things ready. Tom Alexander, a neighbor, said of Mike, "Around and around he goes and where he stops nobody knows." They packed the rumble seat with enough clothes to last for a couple of months. Packed plenty of food. To save money they planned to cook most of their meals. They had some sausage and a lot of fruit. On the day of departure, Mike and Bill posed for some pictures with the vehicle. Uncle Tom Alexander had the last word when he remarked, "Look at him standing there as proud as a dog with a bladder full of piss and a street full of lampposts." They left about 7:00 a.m. and there were several people to see them off. They drove approximately 100 miles that morning and reached Uncle Carl, who was teaching school at Williamson. They planned it that way since they not only wanted to see their uncle, but they wanted a free meal as well. They left after lunch and drove another 75 or 100 miles when the first sign of trouble showed up. They stopped to relieve themselves along the road. When they had finished the old Ford would not cooperate. It was about 5:00 p.m. by this time and they walked to the nearest house. An old man and his wife lived in the very modest four room home with their daughter and her four children. The old man didn't know anything about cars, but invited them to stay the night and got some help in the morning. The old folks had a son in the CCC camps and lived and raised the family on the $25.00 a month their son sent home. They didn't have much to eat and Mike and Bill went to the car and got enough food for everyone. They cooked the eggs and sausage, fried some potatoes. The folks there had some onions and flour to make gravy.

There were only two beds in the house and the old man gave them his bed. Every one was tired and all were in bed at 9:00 p.m. There was no electricity. They lived in abject poverty. A rooster woke Mike up at 5:00 a.m. and the old lady was already up. They took what food was left and fried some apples that Sadie had put in the car just before they had left Huntington.

The old man contacted a friend about a mile away by hitching a ride with a neighbor. The friend knew something about cars. He located the problem quickly and the boys were on their way. Shortly after they crossed the Georgia border into Florida, some con men had set up a business designed to relieve unwary tourists of their money. They had several animals, mostly snakes and other reptiles, positioned along a curved pathway. A man offered to give them a guided tour. The boys didn't look very prosperous, but the scam began after the boys looked at some snakes and possums and a few other animals. They turned a sharp curve and a hidden booth was noticed with a man behind the counter. The man had a board with a bunch of holes drilled in it. The holes were numbered and for each hole there were marbles. Shake the board and the marbles rolled into the holes. If the score added up to a certain amount you won a cash prize. Bill decided he was going to help finance the trip in a big way. He gave the con man a quarter. The man went through an elaborate pep talk and told Bill he couldn't lose next time. Bill gave the man 50 cents and lost again. Another 50 cents and another loss. Mike could see that prosperity was not just around the corner. He pulled Bill away before the con artist could make himself rich. Bill was unhappy because he couldn't hit the mother lode. He had lost most of the money he had brought with him.

The boys drove on to Titusville, Florida where they began to look for quarters. After three or four contacts they reached an agreement with a nice old couple who had some rooms. The boys were extremely lucky in that the old folks were lonesome and had raised a couple of boys themselves. They agreed to rent a room with a bed and bath for $5.00 a week. Remember this was still a depressed economy. The boys snapped it up in a hurry and decided to go fishing. The old man who hadn't been anywhere for months decided to go with them. After that, most everywhere they went the old man was with them. He even helped buy the gas which really helped the cash flow considerably.

Back in those days, oranges and other citrus fruit were in abundance. The boys always took oranges back to the nice old lady. One Saturday the boys took the old man to a rodeo. Mike met a man who was riding the rodeo circuit. He was from Las

Vegas. Mike was pleased to renew the friendship. Mike, Jr. met several men who rode the rodeo circuits while in Nevada. They usually took care of a ranch for part of a year and rode professionally for the rest of the year. Mike, Jr. became a very good friend of Slats Jacobs, a well known cowboy rider. Slats owned a small ranch near the spike camp.

There was a man who sold citrus fruit by the bushel or by the pound. He was just a few minutes from where the boys lived. In order to get attention, he advertised orange juice for ten cents all you could drink. Mike and Bill drank 50 cents worth of juice every time they went by and paid ten cents each for it. The man just laughed about it and enjoyed the boys' company.

After about two weeks, the boys decided to go to Miami. They left early one morning and were stopped by a highway patrolman, who had been observing their West Virginia license plates. The boys were fast running out of money despite their attempts at economy. When they got to Miami they had to face reality. They wouldn't last two days there. Even a modest room cost $5.00 per day. There was no free orange juice, no motherly old lady to cook their fish. They decided to head back to Titusville immediately and get their old room back. After a very pleasant week they had to face reality again. They started home very low on cash. They stopped off at Daytona Beach on the way to Georgia. It was customary for tourists to run their autos on the beach. You could go for miles. They got out on the beach and Bill wanted to look for shells.

When he had finished and started back to the car Mike was feeling like a little diversion and he moved the car about 100 feet north on the beach. Bill ran to catch up and Mike moved the car again. Finally he kept moving just fast enough to stay ahead of Bill. Eventually Bill was running as fast as he could to catch the car. Finally, Mike let him catch up again. Bill got out his knife and said "If you start up again. I'll throw the knife at you." Mike started up again and Bill threw the knife and broke the windshield. They didn't have money to get it fixed so they drove on north without a windshield.

Somewhere in Tennessee they lost the convertible top. It just blew off and they didn't stop to get it. They kept driving in freezing rain and spent a miserable night in a cheap $1.00 a night tourist cabin.

The next day about 2:00 p.m. they arrived home. It was a

pretty day and lots of people were loafing around Niermann's Drug Store. They stopped the car and when they decided to leave, the car would not start for love nor money. Finally, after many attempts to start it, one of Mike's friends towed it away. Mike later sold the vehicle for $30.00. Mike was dead broke, but the economy was looking up and jobs were coming available.

When Mike, Jr. got his first paycheck. He got boxed in and promised to take one of the girls to the movies and Mike, Jr. never did figure out how the news leaked out, but all his buddies were at the bus stop to watch the great departure. They were very vocal and offered a world of suggestions as to how the time would be spent and where they would sit and what time to return home and so on. Mike, Jr. hadn't expected all the attention and was slightly embarrassed by it. This had an impact on Mike's future involvement with the dating game.

Mike got a job as a laborer, paying 40 cents an hour or $3.20 per day. All this time Mike was a good churchgoer and sang in the choir at times. One afternoon they played poker until dark out behind the church. While playing, Mike caught a little garter snake and put it in his pocket. The choir consisted of seven girls and three boys. Mike put the garter snake in the breast pocket of his shirt. The reptile would stick his head out and hiss once in awhile. If the girls were on a high note they would quit singing and almost start screaming. The congregation could not figure out what was wrong, but the people in the choir knew they were snake bit.

Mike accepted a job as a guard with the C&G Railway. He was given a section of track to be responsible for. He was provided with a small building with heat and a good chair near the gate where the yard engine passed through and was told to take over. They gave Mike a 38 cal. special revolver and $5.00 per day. That was more money than his father made. His shoulder bothered him quite a bit and was occasionally dislocating. The C&G said they would take care of the expenses, so Mike decided to get the surgery he needed. They operated on him sometime in January. He had a little trouble with the operation. The wound pulled apart and infection set in. A pretty, young nurse took good care of him and he began to mend. He was in the hospital for about two months. When he was discharged he had to carry a sandbag tied to his arm to get it straightened out. He had a couple of dates with the pretty nurse, but nothing ever came of it. The war was heating up in Europe and Pearl Harbor had already occurred. Mike, Jr. was so anxious and concerned about getting into the service that he lost 15 pounds and was difficult to deal with. When he enlisted he did not get home for almost three years.

About this time young Billy enlisted in the Sea Bees. He was on several islands in the Pacific and served honorably and well. He was transferred to Alaska when he returned from the Pacific. He married a local girl, was discharged, and attended Marshall College on the G.I. Bill. He later divorced and entered the West Virginia State Police in 1948.

"Little Billy"
Casey's younger brother, July 1948

CHAPTER 4

World War II

Mike was extremely upset about Pearl Harbor and felt like he should be in the service. He could have stayed home with his bum shoulder and his guard's job, but he based all his plans on enlisting as soon as the shoulder permitted. He joined the Home Guard, but was unhappy with it. Mike finally enlisted on July 8 and was sent to Fort Thomas, Kentucky. He established friendships with four or five men in his barracks. They were issued clothing and other necessary equipment. The Army tested them mentally and physically. Everyone was given what they called an IQ test. Mike and his friends heard about the tests from some recruits that were in the group ahead of them. The test scores determined what career field the individual would be assigned to. Mike's group decided to each put a quarter in a pool and the highest score won the jackpot. Mike had been under the impression that he was a dull tool where brains were concerned. His school grades were totally unimpressive. He put a quarter in the pot anyway. When the scores were posted Mike's IQ of 123 was much higher than his buddies who were in a range of 94 to 108. Shots were given to all concerned. Mike was in line for shots and was joking with the man ahead of him. The big needle seemed ominous to Mike when it came his turn. He made the mistake of watching the needle. The next thing he knew they were picking him up off the floor. Mike was one of a small group assigned to the Air Corps and sent to Atlantic City.

The train trip to Atlantic City took Mike through his home town. It was about 9:00 a.m. and Mike wanted to surprise his father. He had a bar of soap in his kit. He wrote a short note and wrapped it around the soap. He was well acquainted with the guard at the gate and just before the train passed, he threw the soap out of the window. Mr. Beckett, the guard, picked up the soap and took the note to Mike, Sr.

Upon arrival at Atlantic City, Mike was billeted at the Runnymeade Hotel. The quarters were very nice, two men to a room. The hotel had been a popular resort and was right on the beach and boardwalk. On going up to register, Mike met an individual that was to become a friend for life. His name was

Howard Lewis and he was a former cab driver from Huntington. They talked awhile.

More testing for technical school and more shots. What time Lewis and Mike had left was spent on the beach. Mike met a girl who was from Philadelphia. She was a very attractive young miss, and they spent hours just lounging on the beach talking. Mike made a date to meet her the next weekend, but was shipped to Sioux Falls, ND and never saw or heard from her again. Lewis was also in the shipment and they met a recruit who also became a friend for life. His name was Charles M. Kenney and he was from Cincinnati, OH. All three of them had been assigned to a technical school involving radio operators and mechanics. Sioux Falls was very pleasant in the fall of 1942, but it must have been the coldest place in the world during the winter months! They nicknamed Lewis "Shag" and the name stayed with him. Mike never again called him anything but "Shag". Kenney was a big tall man 6 ft. 4 in. and a former Cincinnati Reds Minor League ballplayer. He was 29 years old and Mike was 22. Sometimes Mike called him "Dad".

The cagey maneuvering by Shag and Kenney got the three of them into the same barracks. Then the two operators got Mike elected barracks chief. They thought Mike would give them favored treatment when it came to work details. They soon found out differently, and it almost caused trouble. But the friendship became an established fact and they were almost always together. The winter months were coming on and the nights were very cold. They had several men assigned as fire wardens. These men had been assigned permanently to the base and had been put on various details. They were a little bitter and didn't like the new students coming in for training. One of the duties of the fire wardens was to see that the two pot belly stoves had plenty of coal and were producing heat.

The students didn't like their fire warden because he deliberately made a lot of noise. He would slam the barracks door at 0100 or 0200 then make some crappy statements and slam the door again and leave. One night he was especially obnoxious and the students threatened to tear off his clothes and throw him out in the snow. He left and went to get the charge of quarters for the night, a young second lieutenant, who was not popular with the students. The young officer came banging into the barracks and shouted in a loud voice "Let's get some windows open, this place smells like a shithouse." Mike, who was down at the other end of the barracks, said in a falsetto voice "I didn't smell a damn thing until you came in." The lieutenant was beside himself with anger and demanded to know who made the last sacrilegious remark. Needless to say, no one told him. He tried to get even by opening several windows and letting cold air into the barracks.

About a week later, Mike was made a student lieutenant colonel in charge of 800 students. One of the first things he negotiated was a code for fire wardens. There were three paratroopers with jump wings attending the school. They were very clannish and did not associate with the other students. A friendship of sorts sprang up between the three paratroopers, Mike, Shag, and Kenney. The paratroopers lived in a barracks by themselves. They always kept their shoes with a deep shine. One night Shag, Kenny, and Mike went to the troopers barracks and thought they would play a prank on them. Shag and Kenney mixed their shoes all up and lined them down the middle of the barracks. They had about 10 pairs of various style shoes. They created a long line. Later Mike and his friends went to a base movie. When they returned to their barracks they found their shoes all over the place. Big spike nails had been driven through the shoe soles into the floor. It was one hell of a job to get them loose.

Everyone was getting along well at school. Some of the students had acquired a code speed of 25 words per minute. Mike was copying 20 wpm, which was about as fast as he could print. Mike wasn't getting along too well with the mechanics involved, but he got by and was given the MOS 2756 Radio Operator and Mechanic. Shag and Kenney were the first to graduate and ship out. It was a sad day for Mike.

Shag and Kenney were in the chow line for the last time. Mike had some duties to see to and was late in getting to the mess hall. He was 10 or 15 men back of them in the chow line. They wanted Mike up with them so they could talk. No one in the line objected and Mike moved up. A disgruntled, washed-out cadet on K.P. duty put his hand on Mike's shoulder and started to move him to the back of the line. It was more than Mike could bear. His buddies were leaving, and now this indignity. Mike floored the washout with one left to the nose, and he yelled like a Banshee. Mike left before the washout could get to the lieutenant in charge of the student mess. Shag talked about that left hook until the day he died. No one in the chow line would admit that they saw what happened or knew who Mike was. Shag shipped out and later went to India. Kenney left and later ended up in Prestwick, Scotland.

Mike finally got his ducks in a row and was shipped to Casper, Wyoming. He was assigned to the 99th Airdrome Squadron as a radio operator. Mike had no trouble getting into the power structure of the squadron. It was a newly formed unit. There were several vacuums to be filled. The Royal Canadian Army training stood Mike in good stead. He was put in charge of the physical education program and trained the troops in close order drill. They were whipped into good shape and Capt. Dvorack, the squadron commander, made him a brand new corporal with two big stripes on his sleeve. Mike was told that he would make sergeant the next month.

A base officer candidate board met to consider applications for commissioning. Two people were chosen, Cpl. Mike Casey and an active duty colonel's son. The colonel's son was transferred to a base squadron for the next class. Casey remained in the 99th and was told he would be transferred later. Mike was lost in the shuffle. He was never sent to OCS (the commander was more concerned about his unit than he was about Mike's future). Mike obtained several excellent recommendations stating that he was officer material, including one from the squadron commander at radio school.

The Base Commander decided to sponsor a track meet between base units. Cpl. Casey selected his team. Most of them were already in good shape and raring to go. Approximately 16 men were selected for the team. Cpl. Casey placed in three events. He was on the winning relay team. He placed second in the 300 yd. dash and won the high jump. The 99th ran away with the meet and scored more points than all the other organizations combined.

The 99th Airdrome Squadron headed for India the early part of September 1943. Life aboard ship soon settled into a routine. Mike gave calisthenics every day and kept the men in pretty good shape. Most of the time was spent sleeping or playing poker. They were on the boat about 35 days. They docked in Australia, but the troops were not allowed to go ashore. Later the ship landed at Hobart in the Tasmanian Islands. It was a quaint place and there were no motor vehicles to be seen. All that was visible were horse-drawn drays and wagons. The Tasmanians exported a large quantity of chocolate, jams and jellies.

The ship reached Bombay the first part of October. Everyone was restricted to the dock area. British guards were manning the gates. Mike was put in charge of the detail handling the officers' luggage. He talked to the guards about his involvement with the Canadian Army and was well-received by the British troops. The officers were allowed to come and go as they pleased but could not leave the dock area. A major who was a medical doctor noticed Mike talking to the British and decided to take advantage of Mike's connections. The major was known as a BTO (big time operator) and used unorthodox means of doing as he pleased. Mike was a little bit like that too and together they planned a way to get off the dock and into Bombay, one of the most exciting cities in the world. As head of the guard detail Mike wore a sidearm and was well-dressed, and had spit shined shoes, all of which was nicely complemented by a deep golden suntan.

Sexually transmitted disease was a continuing problem. Some commanders were not concerned about it, others were very serious about establishing controls. Mike and the major devised a scheme to leave the dock area and inspect the brothels. In the few days it took to leave Bombay and get to Karachi about a dozen men were infected. Mike met two of them while he was in the hospital with amoebic dysentery. The commander in Mike's area was quite concerned and never failed to chair an excellent program. In England the problem was so serious that one prominent British lady recommended that all servicemen returning from India wear a red armband for several months.

The major and Mike, the corporal of the guard, had hatched a plan whereby a four man team of medics, with Mike as guard, would go into town for the express purpose of looking over the brothels and designating the ones acceptable for the use of American troops. All others were to be put off limits. The Brits were impressed and let them through. The officers chipped in and paid rent on a taxi cab for three or four hours. They went to a Chinese restaurant that had been approved for British troops. Mike ordered prawn with fried rice and was never the same again. After eating some of the best food they had ever tasted, the group set out to explore the city of Bombay. No one questioned the major's authority; he was a take charge guy anyway. He was very emphatic on insisting that there be no contact with the prostitutes. After seeing the films on sexually transmitted diseases during the voyage over, all the men were in complete agreement with the major.

The group went to the Grand Hotel for a drink. A beautiful curved stairway led from the lobby to the first floor. Coming down the stairway was a British captain. He carried a swagger stick, and his name was Blackie White. Capt. White saw the Americans and made a beeline for them. Capt. White had been posted to India for six years and he still had one year to go. He was a brilliant, well-informed man and he enjoyed associating with Americans. He was very proficient in speaking three or four dialects of the Indian language. The captain decided to join the Americans, and Mike's group relocated their cab and headed for the Grand Road, an area of town where the most expensive brothels were located. The houses looked like palaces. Capt. White introduced the madam as being Russian. She immediately ordered refreshments. All the officers accepted the drinks on the house, but Mike, who as yet had not experienced any drinks stronger than the British beer he drank in pubs. The group visited several establishments and finally ended up down at the cages. The cages were located in the most disreputable section of town. Hundreds of cheap prostitutes were housed behind bars. Their services could be purchased for about six annas or eight cents American money. VD was commonplace—practically all the prostitutes had it. The cages were a very depressing place and were managed by the city of Bombay. They were operated for the benefit of the lowly paid dock workers.

The group finally said good-bye to Capt. Blackie White and returned to the ship and were transferred to a British troop carrier for the short run to Karachi. It took about three days. It was a miserable and degrading experience, worse than the trip Mike originally made to England. The vessel was extremely crowded and the food service was horrible. Quarters were crowded with four tiers of bunks in a dim, ill-smelling section of the ship. When they landed, Mike had amoebic dysentery and was taken to the hospital immediately and stayed there for three weeks.

The powers that be split up the 99th Airdrome Squadron and sent the personnel to air bases in the Assam Valley. It was a tea growing section of the country and was hot and humid. It was probably a good thing for Mike that he didn't go up the valley with his squadron. Three of the radio operators were killed the first month, including one of Mike's best friends, a young Italian boy that helped take care of Mike's money when he was gambling. The afternoon Mike got out of the hospital, he won about $450.00 playing blackjack. He put $400 of it into a soldiers' deposit at 4% interest.

The very next day, orders came for Mike to be in charge of a 20 man shipment to Calcutta. It was a three day trip almost all across India by train. The 20 men rode in a boxcar with benches to sit on. Mike had a seat with some British sergeants at the end of the car ahead. The men resented the travel accommodations and rightly so. It was a travesty. The train was operated by Anglo-Indians and they operated in the British tradition. Non-commissioned officers always had a seat. So Mike rode in comparative comfort while his charges were miserable, hot, and dusty.

They were given enough K rations to make the trip. Mike decided that the rations would be easier to handle if they were split up among the men. This was taking a chance of some non-conformist eating all his food the first day and starving the rest of the time. The rations were split up and it worked rather well. No one had to carry more than their share. Potable water was a definite problem. All of the troops had water purification pills and full canteens with each man getting an extra canteen. There were a few places along the route where water could be obtained, but no one trusted them.

Every type of human affliction imaginable was seen on the trip. Elephantiasis was not uncommon and many people had a leg and foot that looked like an elephant's leg. Beggars were plentiful. "Buchese, sahib! No mama. No papa." was a common cry among the children. It was impossible not to feel sorry for them. Most of the lower caste chewed betel nut. It's leaves were a mild opiate and stained and discolored the teeth and mouth. Cows were considered sacred and were allowed complete freedom. If one got in the way of a street car or other traffic it was full stop ahead. No one would drive the cow out of the way.

The shipment made it to Calcutta late in the evening. All hands arrived without serious difficulty. One man was running a temperature and was sent to the nearest dispensary. Five men, including Mike, were sent to Dum Dum Air Base about 15 miles

Casey and his beard

from Calcutta. Mike was assigned to the Air Transport Command as a flight radio operator. Mike was housed in a long Basha-type barracks with a thatched roof. The barracks were split up into two-man rooms. Two charpoys, or beds, were provided for each room. The charpoys had rope mattresses. Each of the two rooms had a bath with a sink, commode, and shower in behind the barracks connected by a roof.

While Mike, Jr. was on the briny deep, the yen to grow a beard was irresistible. Mike, Jr. grew a very luxurious beard and kept it trimmed to perfection. He had a picture made in Calcutta wearing one of Green's sun helmets. Mike, Jr. was proud of his beard and resented it when the base commander told him to shave it off, that it would interfere with his oxygen mask while in flight. Mike, Jr. shaved it off. Later in his career he was sent to Japan with the occupation forces and he had his picture painted by a Japanese artist. The picture was painted on silk. Mike, Jr. had a mustache several times in his career, but he never grew another beard.

Insects and lizards were rife and made a lot of noise in the ceiling. There were plenty of snakes in the area. There was some excitement the first morning after the shipment arrived. One of the static chasers, two doors down from Mike, discovered a cobra and was too scared to kill it. By the time everyone got organized, the snake was long gone. The men had no duties except flying and they flew about every other day. Pinochle was the favorite card game, and there were always two or three games going on. When there was a light breeze blowing, it was comfortable on the porch. The flight schedule was always posted on the porch and in town at the police station.

The sounds and smells of India were like no other place in the world. It seemed that one could never get out of sight of a funeral procession. Wood and other fuel was very scarce. The women followed the sacred cows and other beasts to pick up the manure. They would gather it in a sort of apron around their waist and slap it up against a wall or some other structure. It would dry quickly. They used the resulting chipata to cook with. It created a very peculiar but not unpleasant smell. The Indians loved highly seasoned food and used a lot of spices.

Rickshaw wallahs would run all day. Their life expectancy was about 28 years. One day Mike was coming out of the Red Cross building and noticed a body dangling from a tree. A rickshaw was nearby. It seems the rickshaw wallah was slowly starving to death and decided to end it all. He pulled his rickshaw up under the tree, stood up in it, and tied one end of his sari to a limb and the other end around his neck, and kicked the rickshaw out from under himself. No one bothered to cut him down.

White women were a distinct minority in India. There were a few WACS, some Red Cross workers, and a handful of nurses coming in. The enlisted men were at a definite disadvantage in the dating game. Mike had a couple of dinner dates with Anglo-Indian girls. They were very pretty. Both were in the service, but they did not have much in common with Mike and he enjoyed the fried rice more than he did the girls. He had several dates with a Welsh lady who worked at the Army-Navy store. She was three years older than Mike and already had a child.

The Welsh lady introduced Mike, Jr. to an Englishman whose last name was Green. He was about 60 years old and had been in the Burma-India area for about 40 years. He was fluent in the Hindu language and was a refugee from Burma. He owned a small company that produced canteen covers and other things such as sun helmets and swagger sticks. He walked out of Burma just ahead of the Chinese forces and settled in Calcutta. He was still in the same business, but worked for someone else. He had a son, about 15 years old, who developed street smarts and made a living on the streets of Calcutta's black market. The boy's mother had stayed in Burma. Green was a friendly person who liked to cook and play Bridge. He was acquainted with several British Army officers and invited them to his apartment on a regular basis. He used to feed them Burmese dishes and, after a fine meal, had a Bridge game. Mike was a frequent guest and really enjoyed it. Green was a good friend. He said he was never going back to England and he probably never did. Mike and Green went to the horse races several times while in Calcutta. The pleasant times with Green made a big difference in the ten months that Mike was in Calcutta.

There was a co-pilot from Huntington. His name was Tom Downey. Mike went to the races with him several times, and they were becoming good friends. One stormy night, he took off in a C-47 with 23 passengers aboard. The flight was never heard from again. Since the plane was never found, no one has the slightest clue as to what happened, and Mike lost a good friend.

Busses ran a regular schedule between the base and Calcutta. Transportation was not a problem. The city had many famous restaurants including the best of the Chinese. One could get everything one wanted, from shark fin to bird's nest soup. There were many Chinese who made a living gathering bird nests for soup. The birds built nests in cliffs and other dangerous places. The Chinese had long poles to punch the nests down. The birds were some sort of swallow and secreted a substance that was sticky and would bind the twigs together to make nests. The Chinese boiled the nests to get the saliva to make a tasty soup.

Calcutta was a city. About to come apart at the seams. It was estimated there were 10 or 11 million people living in the city. About 25% of them slept on the streets. The life expectancy was about 30 years. It was a fabulous place to shop—always a lot of haggling and horse trading. The street merchants had three categories of prices. One for the Indians, one for the British, and one for the Americans. If one had the patience one could get some real bargains.

One day the first sergeant asked Mike to relieve the corporal of the guard for two hours. Mike put on fresh khakis and reported to his duty assignment. Mike hadn't started smoking yet, but he had a cigar one of the Brits had given him which read, "It's a boy!" Traffic was slack, and Mike lit up the cigar and was puffing on it like a big shot. A colonel and his driver slipped in unnoticed. Smoking was not permitted while on gate duty and the colonel was livid.

"Corporal, what's your name?" the colonel shouted. "Corporal Casey," Mike replied. "What are you doing with that cigar in your mouth?" the colonel roared. "I was just getting ready to light it. sir!" answered Mike. Mike saw the driver try to repress a big laugh with no success. The colonel got out of the car and took Mike over to one side and said, "Are you trying to make me look bad in front of my driver?" He took Mike's name and serial number, but as far as Mike could determine, he never reported the incident.

One night Mike was designated a perimeter guard. He was issued a 45 cal. pistol and a jeep and was told to travel the perimeter at least once per hour. Mike knew he had a dull night ahead and he decided to liven it up a bit. He had a spotlight on the jeep and there were plenty of jackals around. A jackal is a scavenger something like a coyote. He started chasing jackals around the level area of the runway. He killed three jackals that night.

Mike finished his check flights and was put on the schedule as a flight radio operator. His first scheduled flight was almost a disaster. It was a dry load which is not as dangerous as 100 octane gas. The load consisted of one aircraft engine and some six inch metal pipes. The pipes had not been tied down properly and when the plane took off and hit some turbulence, the pipes started to move around, creating an immediate hazard. Fortunately, the weather was good and there was no traffic in the pattern. The plane was able to get back on the runway in nothing flat. With a small adjustment and another tie-down, the plane was able to get airborne. Almost 90% of the scheduled flights out of Dum Dum were in C-46s and C-47s. The C-47 was one of the most reliable aircraft ever built. It did not have the cargo capacity of the C-46, but it moved a lot of material.

There were some real characters among the radio operators: two Jewish men from New York, two Polish men from Detroit, several farm boys from the mid-West, and several Californians and Texans. One man from the state of Alabama and his roommate lived next door to Mike. The two were inseparable. Alabama's roommate was killed during a take-off at Tezpur. Alabama was assigned a new roommate. One week later his second roommate was killed in a flight from Dum Dum to Chabua. The plane was never found. By this time Alabama was beside himself, he thought the room had a hex on it. Alabama had a mongrel dog which he was very fond of, and fed him with scraps from the mess hall. One day, a fuel truck ran over the dog and killed him. Alabama was absolutely incoherent. He went to the flight surgeon and was taken off flying status. He didn't leave his room for three days. They sent him to a rest camp for a week and transferred him to a base in the Assam Valley. Mike lost track of him for several months, but later learned he was killed on a flight over the Hump.

About 0500 one morning Mike went to the flight line for a scheduled flight to Bangalore. He was feeling horrible and felt like he was running a temperature. He checked out a parachute and a 45 cal. pistol and headed for the ready room. He met with the pilot, a 20 year old red-headed youth who had flown with Mike before. Mike mentioned the fact that he felt very bad. The pilot delayed the flight while Mike went to the dispensary for a check up. He had a temperature of 103 and showed some signs of malaria. The flight was canceled temporarily and Mike went to the hospital in Calcutta with malaria fever. He was there for 17 days.

The communications officer received authorization to promote three corporals to the rank of sergeant. There were no corporal stripes up for grabs. Mike was promoted to sergeant. As Mike gained experience he began to develop a reputation for dependability. He was never late for a flight. He kept his equipment in good shape. He hadn't as yet developed the habits of smoking or drinking anything stronger than an occasional beer. Pilots began to request him for some of their flights.

One night Mike's crew was ordered into Agra to pick up some priority mail. Before they could leave, three passengers were added to the manifest: Ben Blue, Ann Sheridan, and Jinx Faulkensburg. Of course they wanted to see the Taj from the air. Ben Blue was seated in the co-pilot's seat after take-off and Ben got an excellent view of the Taj from the air. The air space around the Taj was placed off limits to all aircraft.

One of Mike's favorite places to RON was Bangalore in southern India, not far from Ceylon. They had some cozy crew quarters and probably the best chow in India. The weather was delightful. The runway had a built-in obstacle course. There was a 50 or 60 ft. hump in the runway and the aircraft landed on an uphill grade. The weather was so nice the British officers often left their families there during the monsoons. There were two women there who were famous all over the country. They were referred to as the Queen of Hearts and the Ace of Spades. Everywhere the flight crews gathered, they were there to provide entertainment.

The Air Transport Command began to develop a small airline. Regularly scheduled flights originated at Dum Dum and terminated at New Delhi. Sgt. Mike was selected for the first crew. Later more flights were added and Mike had a good thing going. The aircraft used were C-47s. They were bright aluminum and were never painted. They were called the Silver Fleet. Mike's crew left Dum Dum the same time every evening and flew to New Delhi via Agra and remained overnight (RON). It was hot as blazes in New Delhi and the transient quarters were equipped with fans which moved all across the room. The fan had a pulley system which had to be manually operated. In order to be comfortable someone had to be pulling the rope all the time.

The crew usually arrived at New Delhi about 1000 hours, ate breakfast about 1030 and then tried to get some sleep. Things usually started off well, then it began to get warm. Mike always hired a boy to pull the fan back and forth over his charpoy. Invariably, the boy would go to sleep about 1230 and leave Mike sweltering in the heat. After a futile argument which Mike never won, it would start all over again. The fan boys received one rupee a day which was about as much as a dock worker made. There was no use to fire the fan boy because the next one would probably be worse. Things settled down to a routine that was bordering on being dull. There was one flight to Agra Mike would never forget. A wisdom tooth had been bothering him for days and he considered going to sick bay, but decided to make the flight instead. When they landed at Agra, Mike was running a temperature and was very uncomfortable. The pilot had landed at Agra several times, but had never stayed very long. He wanted to see the Taj Mahal, and this seemed a golden opportunity. Mike was taken to the base hospital and the crew lodged at the transient facilities. The next morning the crew was permitted a pass into town and took off later in the day. They got to see the Taj. Mike was laid up for four days, but he was able to amble around. They had to cut out the wisdom tooth and he was able to visit the Taj Mahal twice. It was very beautiful. Mike also found time one afternoon to visit the burning ghats on the river near old Agra.

In India, a body must be disposed of in just a few hours due to the excessive heat. Agra was a large town, semi-desert, and the wood to burn the bodies was scarce indeed. It seemed funeral processions never stopped. The death rate was very high in the inner cities and especially so during the summer and monsoon season. The funeral procession would come over the hill with four people carrying the body on their shoulders. They used a stretcher-like rig. They put it down beside a funeral pyre which had already been assembled. The pyres were built out of what some called sandalwood and burned with an intense heat. They put the body on the pyre and one man went down to the river and obtained a vessel full of holy water. This water was sprinkled on the body and the pyre was lit. There was a huge stone abutment built out over the river and it was crumbling with age and weather damage.

Mike noticed that families with children's bodies went to the abutment and literally threw the little bodies into the river. The river was full of turtles that had been dining there for hundreds of years. The turtles lost no time in getting rid of the bodies. As they grabbed the children and started thrashing around, the water frothed with a bloody foam. Mike learned that most families were poor and could only afford money to buy wood for the adults, so the children were fed to the turtles.

Two enterprising Indian youths had a unique way of capitalizing on the activity. They would obtain a shinbone or some other small bone and tie it to the end of a rope. They threw the bone out into the river and started slowly pulling the rope in to the shore. Invariably there was a turtle following the human flesh in to the bank. When the turtle got into shallow water, one of the boys would dart in and turn the turtle over on his back and let the visitors take pictures of it. For six annas or one half rupee, the boys would jump into the water feet first. They did a lot of thrashing in the water and the turtles never bothered them.

Sgt. Mike was glad to get the wisdom tooth out and get back to Dum Dum. Another round of promotions and Sergeant Mike Casey was advanced to the grade of Staff Sergeant in July 1943, which, among other good things, meant a substantial raise in pay. It also let the staff sergeant make use of the Top Three Clubs at the various bases, and they have some good poker games at the Top Three Clubs.

On one of the last flights he made to New Delhi, Mike was getting ready for bed in the transient barracks when he heard a lot of noise as if a jungle elephant was coming through. The door slammed, and what looked like a Christmas tree with gaudy decorations began moving down the aisle. Mike said in a loud voice, "Can't a fellow get any sleep around here?" Shag was ready to pick a fight. Yes! It was old Shag whom Mike had not seen for over a year. He wore an A2 Jacket with all the trimmings, including a big pair of wings. He wore Natal boots and a leather flight cap and his arms were full of packages. Mike called him a glamour boy and he recognized Mike's voice. He let out a whoop and pinned Mike to the bunk. Mike put on his clothes and they went to the NCO Club and talked over old times. Both of them had early flights and had to get to bed. Shag made arrangements to get to Dum Dum the next week to spend some time with Mike.

The next week, Mike got a call and Shag advised him he was in town and raring to go. Mike got a weekend pass, and they started out by going to Firpo's Restaurant and getting a sumptuous meal. Mike had used his Canadian connections to get invited to the Sergeants' Club at a large British base. It took prestige plus just to get in. There was dancing, and several girls were present mostly Anglo-Indian WAF. They ended up at a party sponsored by the Calcutta Tramway Company. This was the last time Mike saw Shag until after the war.

The British forces were running into trouble about 75 miles south of Chittagong, Burma. They were running out of food and ammunition. The Japs were between them and India, if they went South they hit more jungle, East they ran into the ocean. So! They decided to dig in and wait to be resupplied. The Brits claimed that they didn't have any aircraft to spare so they called on the Americans for help. The Air Transport Command took a C-47 and crew from each one of the lower Assam Valley air bases and sent them to Chittagong. Altogether five aircraft reported. SSgt. Mike was sent with the crew from Dum Dum. They took their own folding cots and some blankets and mosquito nets and planned to sleep under the airplane.

Chittagong was a small Burmese port city and was known all over the seafaring world as a sin city. The port was located up a large river about five miles. The only white people there were the captain of the port and his family. Some British anti-aircraft crews were there to protect the port from bomb damage. Two or three B-25 squadrons operated from these occasionally. The airfield was about five miles from the town near the coast. The next day two C-47s, including Mike's crew, flew into Chittagong with rations enough to stay a week. Each crew member had a 45 cal. pistol with two boxes of ammunition. The first evening they stayed around the plane. They set up the folding cots and tied the mosquito netting to the airplane with cleverly placed wires and started to prepare supper. They drained the sumps on the engines and got a little gas. They filled a K ration can with dirt and poured a little 100 octane in it. The dirt absorbed the gas and created a fire for several minutes.

They made out very well heating up their K rations over the gasoline fire, washing them down with beer that had been cooled with 100 octane gas. One hundred octane gas, with its rapid rate of evaporation, really cools off the beer in a hurry. They had fruit which they purchased in town. One of the men had a softball with him and they manufactured a bat out of an old 2 x 4 and played softball with the other crews.

The next morning a British captain arrived in an American 6 x 6 truck loaded with supplies to be kicked out of the airplane when they flew over the British troop positions. They also supplied two quartermaster privates to kick out the food and ammo. They took off after a very good briefing. It seems that there was a small stream which curved around in front of the Brits and went on to the ocean. The Japs occupied the curve in the stream between the Brits and Chittagong. The C-47s were to make several passes over the area where the Brits were and kick out the supplies. Sometimes a sudden ground wind sprang up and caused some supplies to be carried over to the Japs. Mike and his crew could see them very casually and systematically pick them up. However, the Brits got the bulk of the supplies and the first mission was considered a success.

The Japs would occasionally fire small arms at the lumbering C-47 and they counted five bullet holes, probably about 30 cal. rounds in the aircraft. Five hundred feet was the recommended altitude so the drift wouldn't carry the supplies over to the Nips. It took about one half hour to fly to the area. It was a clear day, and at 500 feet, they had a good view of both the stranded Brits and the Japs. As the C-47 banked toward the sea at the turn in the river, the kickers started getting busy. They were tied in with safety belts and strong safety lines to keep them from falling out the cargo door. The Japs weren't overly concerned about the presence of unarmed transports. They could be seen moving around under the trees waiting for a stray package from Uncle Sam. The next morning, the crew chief picked up several pieces of scrap iron and said he was going to get a Jap. When the aircraft was adjusting its run at the curve in the stream, he threw several pieces of the scrap metal out of the plane. Chances are it never came near the Japs, but it made the crew chief the aggressor for a change. The crews went into the town of Chittagong and had a real nice meal at the Chinese restaurant. Mike still has the menu. It took four days at two missions per day per plane to get the job done. Many times planes returned with several small caliber bullet holes in the fuselage. They later heard that the British were able to fight their way out and were evacuated by sea, and most of them survived.

SSgt Mike Casey was transferred to Jorhat, India where he made a trip over the Hump almost every day. Jorhat had mostly C-87s, the cargo version of the B-24 and C-109, the tanker version of the B-24. The C-109 had two cargo tanks in the bomb bays and was a very dangerous plane to be involved with. The crews nicknamed them the "C-109 Boom" because so many of them exploded while in flight.

Things changed drastically. No more easy going milk runs in the valley. Most flights out of Jorhat went to Kunming and other bases in Northern China. The Japanese controlled all coastal China and had pushed Chiang Kai-shek and his troops back to the North where they waged war with the Japanese. The 20th Air Force was making daily bomb runs over Japan and SSgt. Mike and his buddies were flying 100 octane gas to them at the Chengtu area where there were seven airfields, some of them fairly close together. Crews had been known to be receiving landing instructions at one base and in the traffic pattern at another base. This could get a little hairy. Some crews were lost this way.

India had very few eggs. China was overflowing with them. Each trip Mike made to China he ate four or five eggs, both scrambled and fried, at the various flight line messes. Mike never did learn why eggs were so plentiful in China and scarce in India. Occasionally Mike would fly a dozen back to India. There was no refrigeration to keep them cool. They never were allowed to go to waste. They were good for mixing Tom and Jerry's and were good to throw once in a while.

A flight usually took about four or five hours depending on weather conditions. Sometimes they would hit head winds of 90 to 100 miles per hour and the ground speed would be 50 or 60 mph, and it seemed as though they weren't getting anywhere. Many aircraft and crews were lost and ran out of gas in situations like this.

One day Mike made a crossing with a Georgia native, Capt. R. L. Cooper, and the assistant operations officer whose name was Simpson. Simpson had some influence in operations, and every time Cooper went out, Simpson tried to have Mike with him. Simpson would usually complete the crew. It seems Mike was their good luck charm.

Mike was in China with another crew, and for some reason or another they decided to stay overnight. They slept in the transient barracks after viewing a western movie. They hit the sack about 2230. Just about midnight several Nip fighter bombers came over with just enough nuisance value to cause an alarm to be sounded. It was a three ball alarm and the Americans were piling out of the barracks and into a ditch behind the barracks. The men, mostly transients who were sleeping in the quarters, would get up at night, when nature called, and use the ditch for what it was not supposed to be used for. You hit the ditch expecting to be protected from flying feces and you landed right in it.

Flying the Hump was like nothing else one could describe. The weather was terrible. Sometimes one had to get to 33,000 or 34,000 feet to get above the weather. At that altitude the cold was intense and dangerous. Cooper and Simpson wore electric heated flying suits. Mike was afraid of them and wore the fur-lined leather equipment. He would survive a short circuit and be warm if he had to bail out. At 33,000 feet, one could spit against the bare aluminum of the aircraft and it hit the metal like a rock. Most of the Humpsters tried to time their bowel movements so they would not occur during a Hump flight. If this unfortunate activity was necessary, a newspaper was unfolded on the floor of the cargo area and one took down his pants and again it was frozen by the time it hit the floor.

It was reported at a briefing that the Japs were operating a third stage gunnery school south of Myitkyina and were sending fighter pilots to the area to make practice firing runs on the unarmed cargo planes. This made an ideal training situation for fledgling pilots.

If they ran into Jap fighter activity, the only thing to do was hunt for a cloud to duck into. The overcast generally ran to 30,000 ft., so it was not hard to find a place to hide. Or they could duck into a valley where the much faster fighter plane could not maneuver as effectively.

A strange situation occurred one night. A C-109 tanker called in an emergency. He was lost in bad weather and bucking a 100 mph headwind. He wasn't getting anywhere and was running out of fuel at a rapid rate. He still had two cargo tanks full of gas in the bomb bay. Apparently, the crew chief was making his first flight in a C-109 and didn't know how to transfer the gas from the cargo tanks to the wing tanks. A quick call located the maintenance officer at the O Club. He was familiar with the C-109 and started giving instructions on how to transfer gas to the wing tanks so it could be used. The crew chief attempted to follow instructions, but evidently made some wrong moves because he was pumping gas out of the wing tanks. They ran out of fuel with several hundred gallons in the cargo tanks. The radio operator had been giving some good position reports and their location was known. Mike learned that they walked out to a native village and were picked up.

Mike had an aldis lamp which he used very little as a transmitter, but it was a perfect foot warmer. It put out a high powered beam of light. When directed at the feet, it was just what the doctor ordered. The aircraft were poorly heated, no heat at all in the bomb bay or cargo tank compartment. Many times, the flight line had people lined up to get their canteens of water flown over the Hump and back. Most times it would be solid ice.

There was always some sort of emergency. It was necessary to give accurate position reports back to operations. They needed this info in case the plane went down. The mountains towered to 23,000 to 25,000 feet. Mt. Everest was even higher, but they seldom ever got that far north. It was not uncommon to hit a down draft and fall 2,000 ft.

About the worst situation you could encounter was a load of Chinese for the flight back to India. They were loaded on the aircraft like sheep with no oxygen masks. Just one free flow mask in case of emergency. They had several slop buckets all down the cargo bay. And when the Chinese got sick they gave undivided attention to the honey buckets. They were not dressed for the cold flight and some actually died on board. The Chinese ate a lot of garlic and it took a week or two to get the airplane aired out. The Chinese were flown to Soukerting, in India, and given some infantry training, then flown back to China.

In China, human life was not important in the general scheme of things. Mike knew of several killings on the flight line from hearsay. One, he was a witness to. Mike and his crew had just flown into Kunming. When they got off the plane, a crew parked alongside to unload the cargo. The driver of the truck brushed against the plane causing a slight dent. The supervisor of the group told the driver to come with him. The supervisor took the driver over behind a shelf of barrels. A shot was heard and pretty soon the supervisor returned alone. Nothing was said. Mike reported it to the operations officer, but evidently no action was taken.

There was a story that was almost legend. Mike had heard it many times. Twenty Chinese soldiers were loaded into a C-87 and were enroute to India for infantry training. They were flying a northern route, probably Fox Route. Myitkyina had not been taken, and it was not safe to fly the southern routes near the air base. An engine conked out on the C-87. They could not maintain altitude. The crew, aided by the Chinese, threw all the equipment out of the plane. Another engine started smoking and something drastic had to be done. The young Chinese lieutenant in charge of the group held a brief conversation with the pilot and told him not to worry. The lieutenant gave an order and the Chinese troops stood up. He ordered the men to remove the cargo door. The Chinese lieutenant ordered them to march and the first man in line stepped out of the plane and all the others followed. The lieutenant was the last man out. With the plane lightened by several thousand pounds, they managed to maintain altitude. It is possible that it occurred. The Chinese are very disciplined troops.

Mike had heard of other crews seeing static electricity in the aircraft. One night he had a hair-raising experience. What looked like a ball of fire rolled all the way down the cargo floor in the C-87 aircraft and went out the rear of the plane. He had witnessed the phenomenon of St. Elmo's fire.

Before the Myitkyina Campaign the ATC craft had to fly north to keep away from Jap interference and attack. The farther south one could fly, the lower the elevation of the mountains became, therefore, compromising the safety of the aircraft. The Chinese and British and Americans must take Myitkyina. Mike made two or three flights to Myitkyina, taking ammunition and supplies in and flying wounded soldiers out. The Japs owned one end of the runway and allied forces occupied the other. For several days the Japs occupied a fighter strip nearby while the U.S. Air Force used the main runway. The entire crew was issued 45 cal. automatics.

The radio transmitter was an ancient BC-375-D. The same equipment they used in some tanks. It had a trailing wire antenna that if properly loaded gave one all the range necessary.

Sometimes a radioman would forget to wind in the antenna, and when they landed there was no weight on the trailing end of the wire. Without a weight on the end, an antenna would not reel out if a operator failed to reel the antenna in before landing. He would lose it if no one noticed the weight was missing. The next operator would be without a trailing wire antenna which could be dangerous. This may have left them at a disadvantage in an emergency. Many times they relayed position reports and other information through another aircraft in a better location. Mike always made it a habit of dropping the weight on the end of the antenna to the ground and reeling it back up before a flight.

One afternoon, they were headed for Kunming, China in a C-87. The weather was borderline. The cargo was 100 octane gas in 50 gallon drums, headed for the Chengtu area. At about the halfway point, a little over two hours out, the right inboard engine began cutting out. Later it developed an oil leak. The pilot, R. L. Cooper, was afraid of fire and feathered the engine. They were in no immediate danger, but it would be difficult to maintain altitude. Jap fighters in the area would have a definite advantage. A large four-engine plane loaded with gasoline barrels is not easily maneuvered, especially if an engine is out.

Mike gave a position report and requested clearance to Yunnanye. Yunnanye was an air base that was used mainly for emergencies. It was surrounded on three sides by mountain ranges with only the approach relatively clear. Once on the downwind leg you were committed. There was not enough room to go around and make a fresh approach.

They were about 30 minutes out from the landing strip and the weather was getting worse. It was about 1500 hours and there was plenty of daylight left. Mike transmitted position reports and status of the plane every ten minutes. The pilot was permitted to make a straight in approach.

They were really sweating it out because fighter activity had been reported in the area. But the lack of local traffic was reassuring and they felt good about it. The plane landed without any further problems and was met by all the emergency vehicles on the base. The crew had never been to Yunnanye and probably would not have visited there if it had not been for the engine failure. They flew out the next day on a shuttle flight to Kunming, China. They arrived at their base, Jorhat, late in the evening.

Later in the month, Mike received another stripe and the title of Technical Sergeant. He asked his pilot R. L. Cooper if he had anything to do with it. Cooper denied any knowledge of the promotion.

There was a thriving black market in China. If an airman was willing to take a chance on getting caught he could sell most anything, including used toothbrushes. Chinese coolies would come out to the plane with a big roll of American money and offer to buy articles off the crews. Of course, these coolies were operating for some one else. They would pay $25.00 for a carton of cigarettes. Two or three dollars for a used toothbrush. Several Americans were arrested for black market activities.

By this time Mike had accumulated 1175 flying hours, far more than the average static chaser, and was getting weary and wasn't sleeping well. The crews were given a two ounce shot of whiskey after each combat flight. A trip over the Hump was considered a combat flight. Part of the citation for Mike's Distinguished Flying Cross, which was later awarded to him, was as follows, "Flying at night as well as by day . . . where enemy interception and attack was probable and expected." Mike had been giving his combat ration of two ounces to his buddies. One night the crew came in worn out by a long and difficult night flight. They were debriefed and given a chit to take to the dispensary, indicating that they had completed a combat mission. Mike had not been sleeping well and decided to take his ration. One of the crew had a can of beer and he let Mike use part of it as a chaser. Mike drank his ration, chased it with the beer and staggered off to his basha and never slept better in his life. Mike always accepted his combat ration after that. The crew would sit around for 30 to 45 minutes and savor the beer and whiskey.

Most of the crew smoked and they all lit up cigarettes and enjoyed them. One night Mike came in from a flight. He had a bad cold when he left, and it was not any better on his return. His fur-lined jacket was covered with what seemed like a pint of frozen mucous around the collar. The crew had their ration and lit up cigarettes. Mike lit one up, too. That was when he started smoking at age 22.

After several more gut-wrenching flights, Mike was recommended for the Air Medal, and the Distinguished Flying Cross. He already had a Presidential Citation and three Battle Stars. Before the orders came through on the decorations, he was sent to a rest camp. The camp was on a British tea plantation and had some nice quarters and other facilities. The camp was run by a technical sergeant from the Army. He was a specialist and knew about hunting and fishing. The chow was good. The atmosphere was good. The camp was located at the base of a mountain, and during the monsoon a kayak ride down the mountain stream was a real treat. The mountain stream ran into a large river and Mike saw his first river dolphin. The river dolphins were very scarce and Mike never saw another one.

Mike had a hair-raising experience at the rest camp. He went bird hunting one morning and was driving a jeep. Mike killed several game birds and was driving back to the camp when a native Indian ran out on the road yelling at the top of his voice, "Jungly elephant, Sahib." Mike stopped for a second and heard the rogue elephant tearing down basha huts and making terrifying noises. Mike got the Indian into the jeep and backed up to where he could see the elephant. He was destroying everything in his path. A rogue elephant almost always has to be killed. Mike put the pedal to the metal and got out of there in a hurry.

Everyone wanted to get a Bengal tiger, but no one did. The food was good and you slept all you wanted. It was the first week off Mike had in two years. He was sent back to Jorhat where, lo and behold, his orders to return stateside awaited him. He cleared the base in nothing flat. And his buddies gathered together enough Jingbough Juice to throw a party. Jingbough Juice is a rice wine with which they run some automobiles in China. The stuff was so awful that they had to disguise it by making a batter and serving it as Tom and Jerry's. Cooper and Simpson came to the party and they were genuinely upset about losing Mike as a friend and fellow crew member.

Mike's orders called for him to go to Karachi and get in line for transportation to the United States. He hitched a ride to Calcutta and stayed at Dum Dum overnight. Not too many of the old crowd were still there. They had all heard of Mike and they gathered at the enlisted men's club to greet him. Mike went into Calcutta and contracted with a tailor to make him a set of suntans out of material like what the officers wore. It was real flashy with blue cloth wings trimmed in white, a name plate sewn over the breast pocket, and a designer China-Burma-India (CBI) patch with silver thread. The tailor had it ready by the next day. He must have worked all night. It was a flashy outfit. He already had a pair of half boots made up for him near Jorhat. He arrived at Karachi and felt lost. He was just beginning to feel the excitement of going home.

The next day he was on orders to board a C-87 for Accra on the Gold Coast of Africa. Scheduled stop-over for fuel was Aden, Arabia. Upon arrival at Accra, Mike was bumped by a lieutenant colonel, Mike was not disappointed. He did not object to spending a few more days in Africa. Mike rested a whole lot and went swimming in the ocean every day. He watched the Africans take their nets out every day and circle back in a huge arc and pull the nets in by hand. Quite often they made a good catch. They sang in cadence as they pulled the nets. It was something to hear. Mike won about $300.00 in a poker game while there. After several days Mike was picked up on a space available basis and boarded a C-87 for Natal, Brazil. There was a stop to refuel at the Ascension Islands. The island was like a rock lost in the middle of the ocean and it takes some skill to hit it on the head. A navigator was always used on this leg of the flight. Thousands of birds used the rock as a mating ground. Those fair feathered friends made it unsafe for aircraft taking off. They refueled and left for Natal. They remained overnight at Natal.

They took off the next morning for Miami, Florida with stop overs at British Guinea and Borinquen Field, Puerto Rico. Upon arrival at Miami, all aboard were checked by customs and given tickets to their home of record. The trip to Huntington took one and a half days. Most of the family was at the train station to greet Mike. He got off the train looking good until his mother saw a cigarette in his hand. She didn't say hello or I'm glad to see you. She said, "What are you doing with that cigarette in your hand?" Mike started to say, "I'm just getting ready to light it, sir", but restrained himself. He was remembering the colonel at the gate in Calcutta at Dum Dum. Sadie was beside herself and said, "There is my big black angel." Mike did have a deep tan which showed up well, but he certainly wasn't an angel.

During this time Ted Smith came home after being in a prisoner of war camp for several months. Ted and Mike ran around together. They met two attractive girls and were invited to a party at their house. Men were still scarce and two more men was an asset not to be ignored. Mike attached himself to the girl who had sponsored the party. She was a very attractive girl with a bubbling personality. She fascinated Mike with her intelligence and wit. Mike met her family and liked them right away. Mike dated her often while home and wrote to her occasionally when away. They made no promises and felt free to date other people, in other words, it was a "no strings attached" relationship. The furlough wasn't going well. Mike was used to making his own decisions and doing whatever he wished within reason. He had plenty of money and wanted to do something interesting and perhaps exciting.

Mike finally decided to go to ATC HQ in Cincinnati a few days early. He told his mother that they had an assignment for him. He arranged transportation and left that afternoon. He arrived at the railroad station about 6:00 p.m. There was a section where service men could go for assistance if they needed help. Mike wanted a decent hotel room since he intended staying three or four days in Cincinnati. He walked up to the young lady who was in charge and told her what he wanted. She asked how much could he afford. He promptly replied $5.00, which was a lot of money in those days. She said there was a fine hotel with some rooms available and that she had some connections there. The price of the rooms was $4.00 plus tax. They talked a little about Mike's Air Force career and she fixed him a cup of coffee and gave him a doughnut.

Mike went to the hotel and it was great. He got a call from the lady that helped him get the room. She said she just wanted to find out how he was doing. Mike told her how nice the room was and how much he appreciated it. She talked for several minutes and she asked him some personal questions about his service and his family. Mike was enjoying talking to her, but he never once thought of making a pass at her. He thought she was too nice to really get involved with a stranger. After they said good-bye, Mike began to put their conversation in its proper prospective and realized he had probably passed making contact with an attractive female. He decided to be more aggressive the next time. Mike slept until 10:00 a.m. the next day and went bowling. After the bowling he caught a late afternoon movie. About six he went to a restaurant that was supposed to serve the best chili in Cincinnati. While enjoying the chili Mike struck up a conversation with two girls in the next booth. The girls lived in a housing unit owned by a church group. They had strict hours and if they were out past a certain time they could not get to their quarters without some embarrassment and hassle.

Mike was in a good mood and asked the girls if they wanted to meet with him later that evening. The girls apparently had talked it over and decided it would work better one on one. And the oldest girl (about 25 years old) stated that she was tied up for the evening and couldn't go out. Mike arranged to meet the other girl at 8:00 p.m. near where she lived. They were both attractive girls and Mike was not too concerned about who he would date. He let them make the decision. They went to a service club near the railway station and sat and talked for about a half hour. The girl, a healthy blond, let it be known that she did not object to an occasional beer. They went to a nice, quiet tavern that she had visited before.

They had three or four beers and Mike, knowing they had a ballroom at the hotel, asked her if she wanted to go to his hotel and dance. The girl whose first name was Margaret (everyone called her Marge) agreed. She was impressed by the atmosphere at the really nice hotel. They danced for a little while and Mike asked her if she would like to see his room. She said that she would like to see his room, if Mike would promise not to make a pass at her. Mike had a lot to learn and he promised her he would conduct himself properly. They went to his room and Mike didn't lay a hand on her. He found out that she had a brother in the service. She had lived near Columbus before the war and went to Cincinnati to get a job. Mike took her home about 11:00 p.m. and didn't even kiss her good night.

The next morning about 8:30, Mike was shaving and the phone rang and Marge was on the line. She stated that she and her girl friend had a talk over coffee and she told Marge she was crazy to be working in a factory with a nice friend like him in town. Mike agreed with her. "What can I do about it?" she said. "You can get the rest of the day off and spend it with me." replied Mike getting braver. "Let me talk to my boss," she said. Mike agreed with her suggestion. About 20 minutes later here she came knocking on Mike's door. She had apparently arranged with her boss to get the day off before she called. They spent the rest of the morning getting acquainted and went out for lunch. They went bowling in the afternoon and out to dinner in the evening. Mike really liked her, yet he knew he was going to Memphis the next day and would probably never see her again. Mike called Kenney's parents and enjoyed talking to them. They were expecting him home in a few weeks. Mike was beginning to wish he could stay for a few more days, but this was not to be.

CHAPTER 5

Post War

Headquarters for the Air Transport Command was in Cincinnati and Mike checked with them about every day. He learned that he was to be stationed at Memphis, Tennessee with the Ferry Command. He asked for an extra day in Cincinnati and it was granted. The girl at the reception desk in ATC HQ let it be known that she was available. Mike was puzzled. He had received more attention from females in the last week than all the rest of his life. Different women had indicated an interest in him. He decided to make hay while the sun was shining and on a very modest scale he did just that.

After an uneventful trip by train to Memphis. Mike was assigned as a flight radio operator. Technical sergeants were very common at the base. Many operators who had served with the Air Force in Europe were there and some from the 8th Air Force, all combat veterans. All that was required to remain on flying status was four flying hours per month. Many of the operators did just that—four hours per month. Mike visited the squadron orderly room to get acquainted and to try to get some good foreign ferry missions. On his third night at the base, Mike was at loose ends and he visited the Noncommissioned Officers' Club where he had already stored a fifth of good whiskey. It was an outstanding club and Mike had a bottle stored in a special room where he could get it any time he wanted it. Mike had not had time to get acquainted and he felt a little lonesome. He selected a table at an outside location near the sidewalk. There was a three foot stone fence surrounding the patio to keep people from coming in from the street. Mike wasn't much of a drinker and did not indulge too often. It was about 2000 hours and quite a few people were walking back and forth where Mike was sitting. He struck up a conversation with a staff sergeant who had been in India. About this time a bouncy WAC came up the sidewalk. She spoke to Mike over the fence and went on up the street. The staff sergeant left and the WAC came back down the sidewalk. She stopped and asked if Mike was new at the base. He indicated yes and she said that she had to get back to her duty station. She was working with a link trainer which is a cockpit of a plane used for training pilots. It seemed they had so many pilots they ran the link trainer until midnight on certain nights.

About 2130 hours she was back and stated that things were a little slack and she liked to walk. Mike asked if she would like to take a walk with him when she finished her tour of duty. She said she would get off duty at 2200 and would be happy to take a walk with him. Mike asked if she would like a mixed drink. She nodded her head and Mike mixed her a drink. She was back shortly with an empty cup. She was a very pleasant person and this time she talked for about five minutes. Mike mixed her another drink and she left. Just before 2200 she came back and said she was finished for the day. She was a private first class and not allowed in the club because of her rank. Her name was Cain and Mike called her Sugar Cain. They went up the street to a softball diamond with bleacher seats and sat and got acquainted with one another. A little after midnight, Mike took her to the small bridge that crossed a creek and stopped there and watched her until she got to her barracks. The area was out of bounds for male personnel. By this time it was almost 0100 and Sugar Cain was supposed to have been in her room at midnight. Mike went to his quarters and hit the sack.

The next morning about 1100 hours, Mike was walking down the street by the link trainer building and saw the Killer washing windows. She acted mad and said she had received extra duty washing windows because Mike had caused her to be late for bed check. She didn't stay mad long and they went to the base movie theater that night.

In the meantime orders came through for Mike to participate in a ferry mission to England. It took a lot of doing on Mike's part to get this prize. He had been trying to get back to England ever since he had left. He managed to sign on as an assistant radio operator for the flight. They went to Hunter Field, Georgia to pick up a brand new B-17 loaded for bear. It was scheduled for a base near London, and London was near Oxford and Margo was at Oxford. The first leg was Presque Isle, Maine. The crew spent a pleasant day and night there waiting for improvement in the weather.

From Maine they went to Goose Bay and on to Greenland. Greenland was a dangerous place to land. Pilots did not elect to land there unless it was some type of emergency. After leaving Greenland, they dropped in at Preswick, Scotland. Mike managed to locate his old friend Charles Kenney who was the tower operator at Preswick. The flight terminated at an air base near London. They obtained quarters in London near Hyde Park. That evening operations contacted the pilot and advised him that the crew would fly a war weary B-24 back to the States. This meant another three days in London which Mike thoroughly approved of. But! It also meant that the crew would be risking their lives to get some junk back to the States. Many, many B-24s and B-17s were flown back to Willow Run, Detroit and scrapped. No one ever put a number on it, but several crews were lost due to this activity and Greenland realized some emergency traffic from it.

Mike traveled to Oxford the next morning and learned that Margo was teaching in Australia. Her sister had married a soldier returned from India and they were still at Oxford. Mike did not try to contact Margo's sister. He returned to London and met a registered nurse at Hyde Park and had tea with her. The nurse was a very sharp individual and knew London very well. She escorted Mike to some very interesting places. The nurse went to work on the evening shift and Mike met her the next day. They had brunch and spent the day together. The crew was leaving the next day and taking a pile of junk with them. The nurse went with Mike to the air field to say good-bye and there was much speculation among the crew members as to Mike's activity while in London.

The war weary B-24 was something else. It had been in use too long. It actually had 2 x 6 inch braces to keep the waist from collapsing. It was not fit to fly the Atlantic Ocean. Operations insisted that they land at Greenland because of the uncertain condition of the aircraft. There was much salvage value to the aircraft. Many gallons of fluid such as hydraulic fluid, oil, de-icer, etc. The fact that they took the junk to Willow Run, Detroit indicated it would become recycled automobile parts. They took off for Greenland. To get to the air base, they flew several miles up a fjord and banked to the right to get on the down wind leg. At that point they were pretty much committed to landing, especially when flying a war weary. There was quite an aircraft graveyard at the upper end of the strip.

The war weary was doing OK. The braces were doing their job in the waist. The two right engines were developing oil leaks. The crew chief did an outstanding job in getting some

Casey is awarded the Distinguished Flying Cross.

maintenance people to take a look at it and make some repairs. They took off the next morning for Presque Isle, Maine. The junk was delivered to the industrial interests. The crew felt that they had been unduly exposed, but lucky.

Mike's Air Medal and Distinguished Flying Cross finally caught up with him. The presentation was set for the first part of June. The base commander, Col. Richardson, made the presentation with all the hoopla and picture taking. The information and pictures were sent to Huntington. Mike, Sr. and Sadie were very proud.

After returning from a foreign ferry mission, the crew was given a three day pass that did not count against leave time. Mike usually stayed on the base, swimming and just plain loafing. He still dated the Killer. She had long since forgiven him for getting her busted from the lofty rank of PFC.

One night Mike was swimming at a civilian-run pool outside the gate. They were having a party. Tubs full of beer and lots of potato chips and dip were available. Mike was invited to join the crowd, most of whom were from the base. Mike accepted a beer and struck up a conversation with an unescorted WAC. She was a little top-heavy and wore a tight tee-shirt. She left Mike and went swimming. When she got out of the pool the tee-shirt was really clinging to her ample bosom. She walked back over and started talking to Mike. A half-lit crew chief walked over and started fondling what the tee-shirt failed to cover up. Mike wasn't with the WAC or in any way responsible for her presence, but he felt it was an insult to him. Besides he didn't get along too well with crew chiefs. One of them had called him a glamour boy because of his blue wings and tailored suntans. Mike hit the crew chief square on the chin, knocking him into a tub of iced beer. Someone had to help him out of the tub. The WAC was quite indignant about the incident. But! She thanked Mike profusely. By this time Mike was disgusted with himself and the rest of the crowd. He left the area, and went to find Sugar Cain.

The next trip of any significance was to Biak, New Guinea about 20 Aug 45. The crew flew to Long Beach, California. They picked up a B-24 and flew it to its final destination. It was a long, tiring trip. The aircraft was in excellent condition. Authorities at Biak indicated that the aircraft was badly needed for air sea rescue. The crew flew back to Honolulu in a C-54. After a two day lay-over they returned to Memphis. At Memphis another three day pass awaited. While on the three day pass, Mike met a civilian lady at a USO dance. He saw her three or four times and liked her very much. She had a two year old daughter. The baby's father was an Air Force navigator who left a very nice girl to face the music alone. Mike thought about getting better acquainted with her, but he decided that he wasn't broad-minded enough to be a good father and husband which she so rightly deserved.

In early September, Mike was with a crew that ferried a C-47 to Natal, Brazil. The aircraft had seen duty with the Alaskan Command and was in good shape. The crew flew to Nashville by commercial air and picked up the aircraft. They spent two days in Natal and Mike had another pair of Natal boots tailor-made and lined with the best leather available. When Mike returned to Memphis, he learned that Killer Cain had been transferred to Wright Patterson AFB to help process the returning service men. The war was winding down. Mike applied for duty with the Arabians. It seemed that the Air Force was committed to supply the Arabs with a new C-47 and crew for diplomatic and political reasons. He also applied for a position with the Philippines mission. The position with the Philippines never materialized and the crew space for the Arabians was given to a master sergeant. Mike had another three day pass.

The 8th Air Force was out-processing flight crews every day and there were so many around that foreign ferry missions were hard to come by. Mike decided to try civilian life again. He applied for a discharge on or about 23 October. Mike was shipped to Wright Patterson. The first person he saw was Killer Cain on a flight line bicycle. She had access to all lists of personnel for separation. She managed to get Mike on her work sheets. She had applied for leave several weeks back and it was granted. She took leave the next day after Mike arrived and Mike never saw her again.

Mike enlisted in the Reserves as a technical sergeant and was discharged. He hitched a ride home with a staff sergeant who lived in Barboursville. Mike had not been a civilian for one week when he found out that he had made a mistake. He enrolled at Marshall College as a journalism major. He spent most of his time at the student union building, playing bridge and hearts and made some money at it. His grades were not impressive and at the end of the semester he decided to re-enlist. He learned that

Casey, Communications Chief,
Third Emergency Rescue Squadron, Japan

This aircraft was picked up in Japan and flown to the
3rd Emergency Rescue Squadron and then taken to Manila.

Last flight of the "Cat" The crew left her at Tokyo
Casey first on left.

he would lose a stripe for being out longer than 90 days. He decided to re-enlist anyway. The loss of a stripe really hurt. It made a big difference in pay, including flight pay. In the enlisted rank structure technical sergeant was a highly prized rank for a flight radio operator.

Mike re-enlisted and was shipped to a depot in South Carolina. Three days later he was shipped to Osaka, Japan and two or three days later was assigned to the 3rd Emergency Rescue Squadron in Itazuki, Japan. The 3rd had many pieces of Air Sea Rescue equipment. They had three B-17s with 40 ft. boats on the bottom of the planes (the boats could be dropped by parachute), one Catalina Flying boat, one L-5 Liaison aircraft, and six high speed Air Sea Rescue boats. The 3rd Emergency Rescue Squadron had flights at four or five different locations in Japan. Korea was covered by a Flying Dutchman and crews from the other flights on a rotation basis. Mike's flight spent one week out of every month at Kempo AFB, Korea.

They had a good runway at Itazuki, but living quarters were scarce. The flight was housed in a aircraft factory until they could put up some tents near the aircraft. It took about three weeks to get tent city in shape for occupancy. The top three grades of enlisted men lived two to a tent. The rest of the men lived four to a tent. A large tent was set up as a dining hall. The dining facility could seat about 75 persons at one sitting.

Mike was the ranking NCO in the Communications Section. His title was Communications Chief of the Unit. The Communications Section was housed in a Quonset hut. There were two other radio men, one PFC, and one corporal. Neither had the experience that Mike had been exposed to. Several Japanese ex-service men worked for the Occupation Forces. Many of them had held high rank. Mike had a man working at crash boats that had been a high-ranking naval officer. The 3rd Emergency Rescue Unit flew surveillance missions about once a week. Sometimes they would fly within sight of Vladivostok. The mission was to chart and report shipping in the area. They reported the type of ship, the location, direction of travel, possible cargo, and speed. The Russians never tried to interfere with the flights, but there were some tense moments and intimidating language. The Russians didn't like the Americans so close to their soil.

Mike made a trip to the rescue flight location at Hokkaido.

One of the communications people gave him a small puppy. The little mutt would fit easily in one hand. Mike was pleased to take the little ball of fur home as a pet. They named her Poncho.

The men made her a nice little bed in the Communications Section. She was spoiled rotten. The second day she sat in some battery acid that had been spilled on the floor and got acid all over her behind. She started to drag her butt across the floor trying to get some relief. Mike picked her up and put her in a pan of cool, clean water and rinsed her posterior off. Finally, she settled down. When she was five or six months old they let her follow the truck from the Communications Section to the flight line. If she noticed a Jap in the area, she would give him a buzz job.

Mike remained in the rescue squadron for about 14 months. The highlight of the assignment was a trip to Honolulu to pick up a B-17. The B-17 had been assigned to a four star general as his personal airplane and was in excellent shape. The hardware to carry a 40 ft. boat had to be installed. Seven people were assigned to bring her back to Japan and Mike was one of the seven. The crew flew to Honolulu in a C-54 and were given individual quarters on the base. The crew ended up spending 43 days in the islands. The Black Widows kept having maintenance problems and the pilots were dragging their feet. The B-17 was also plagued with minor problems. The crew had absolutely nothing to do during this time. They took the B-17 up about every other day. The crew got acquainted with a widow lady that owned the Brooklyn Hotel. Her husband had been a retired Army officer who had spent some time in Japan before the war. She was about 70 years old and owned several pieces of property in the islands. Her husband had helped pioneer air sea rescue tactics and equipment before the war. The crew spent much time at the hotel. They flew down to Hilo and purchased steaks every once in awhile. They knew some girls that worked in the parachute rigging building. The girls usually cooked the steaks. Mike spent all his money and sent home for a loan. Swimming, fishing, laying in the sun, that was the order of the day.

The squadron back in Japan was anxious about getting the plane and crew back in Japan, but the orders were to wait for the Black Widows. They did not have a navigator and Mike's B-17 had two. Another break was the fact that the Black Widows did not have the range of a B-17 and they would have to island

Casey attending an Air Rescue Conference at Manilla.
The entire crew was issued sidearms
as Japanese troops were still in the area.

Famous Marine flagpole, Iwo Jima
Bereman, a crewmember, at the monument, Casey on the right

hop which meant visiting several islands on the way home. The plan was Johnston Island, Midway, Guam, Enewetak, Iwo Jima, and not necessarily in that order. The word came down that they were leaving the next day. All of the four planes had good take-offs and the Black Widows took up a formation and the best air speed was determined. They headed for Johnston Island. Johnston Island is mostly manmade, dredged up out of the ocean.

Midway, where they RON, had some very interesting birds called Gooney birds. A Gooney bird sits on one big egg. The bird is so bulky that it runs to get up take-off speed. The Navy had an air sea rescue aircraft. It was something Mike and his buddies had never seen, a B-24 aircraft with a single tail fin. They used a crew of seven, including scanners. The Navy went all out to feed and entertain the Air Force. The next morning the Navy crew arose early to have breakfast with the 3rd Emergency crew. The Navy served baked beans as an important part of their breakfast. The Black Widow pilots were feeling their baked beans and put on quite a show when they took-off. They made fair time and got to Iwo Jima and decided to RON. The night fighters were still with the B-17, but they were ready to strike out on their own. It was just a short hop to Japan. When the "17" arrived at the air base, there were a lot of strange faces waiting for them. The squadron was now at full strength. Poncho was growing like a weed and was still buzzing the Japs.

The squadron was to send a crew to Manila to attend a rescue conference. They decided to show off the new B-17. Mike went along as radio operator for the B-17 and as a participant in the communications part of the conference. They were gone for one week and Mike got to see Okinawa and the Philippines.

The entire crew were issued 45 cal. automatics. There were still pockets of Japanese troops in the mountains. The Manilla Hotel, which had been damaged during the war, still showed evidence of damage. Mike and his crew stayed at the hotel. Mike kept the receipt showing the payment of three nights lodging. It was said that General MacArthur had an interest in the hotel.

Mike's enlistment had another 18 months to go and he requested Wright Patterson as his next assignment. Mike made another ocean crossing by boat—that made five crossings by boat and probably 20 by air.

The trip from the West Coast was by rail and the train went through Las Vegas and Moapa, Nevada where Mike served in the

CCC program. Mike was assigned to Wright Patterson AFB—still on flying status. Still unable to get his stripe back. Wright Pat was only 175 miles from Huntington and Mike was usually able to hitch a ride. Mike had some money and he was getting tired of hitchhiking so he began to think of a buying vehicle.

These were exciting times in the Air Force. A West Virginian named Chuck Yeager was beginning to make aviation history at Muroc AFB, California. This resulted in many new advances in equipment. Some type of testing was going on constantly. Mike was right in the middle of all the activity. He made three trips to Muroc. On one trip they were doing some testing on a new camera that supposedly could get a picture of a small object on the ground from several thousand feet altitude. He went to Alaska with a crew for cold weather tests in a B-29. They tested various fuel mixtures while in Fairbanks.

Mike accepted a Reserve commission as a second lieutenant. The commission gave him a chance for a new career as an officer. Mike had already passed the board for OCS back at Casper, Wyoming. If they had sent him to OCS at that time he would probably be a captain by now. However, now the top enlisted ranks were full of ex-officers who had re-enlisted. It was very discouraging trying for a top three grade promotion. Mike was sworn in on his 27th birthday as a shave tail. Mike was determined to burn the midnight oil and make up for lost time.

The Armed Forces were full of interesting characters. Benny Brown from Alabama would fit that description. He had been born in the early 20s and had grown up during the Great Depression. He had never thrown anything away. He had washed cars and shined shoes to make an extra buck. Benny had often spoken of the lean days down South during that period. He had said that ground hogs and greens were his favorite food. He had observed that during the Depression a Jack rabbit looked as big as a 100 lb. hog. Benny had been a ground pounder during the war and had been transferred to the Air Force because he had been in the right place at the right time. He had sworn he would never get out of the service. He had never wanted rank higher than a corporal. Mike had gotten pretty well-acquainted with him, and learned later that he owned a prosperous catfish farm in one of the southern states.

Mike was one of the crew members when they brought the first B-36 from Fort Worth, Texas. It was the first one the Air

Force accepted. The number was 005. The B-36 was indeed a huge aircraft. It was needed in the inventory for a specific mission. The first hydrogen bombs were large, heavy, and crude. A B-29 could not deliver them, so they developed a craft with six pusher type engines and the capacity to carry almost a tank car of gasoline. It had a tunnel connecting the cockpit with the waist. There was a small sled on rails that permitted the crew members to go back and forth in the tunnel. The forward momentum of the airplane drove the sled back to the waist. When a crew member wanted to go forward he pulled himself hand over hand on a rope strung along the top of the tunnel. The plane even had a steering wheel to the left of the pilot's seat to facilitate steering on the ground. The B-36s were so big and cumbersome that only a few runways in the country could handle them. Their weight when loaded was tremendous.

One night after purchasing a new car, Mike met his future wife, Martha. Mike had been to the gym lifting weights. He decided to go to the Gatehouse Restaurant for a bowl of chili. Martha was having a late dinner with a friend. He took them home in his new car. Mike asked for her phone number. She either gave him the wrong number or he wrote it down wrong. He had trouble getting in touch with her.

Mike had been going to Huntington about every weekend and visiting his girl friend and her family. Mike told the girl friend that he wouldn't be coming home as often and that he had found a girl that he was interested in. His Huntington friend said she wasn't worried about it and for Mike to do as he pleased. Mike began seeing Martha about every night.

On 9 Sep 48 Mike was ordered to Griffiss AFB in New York. According to paragraph three of the orders, this movement constituted a permanent change of station. The orders called for ferrying three C-47 aircraft to Burtonwood, Lancastershire, England. The C-47s did not have the range required for some of the hops and was equipped with auxiliary gas tanks just behind the radio compartment. The radio went out about halfway through the last leg of the flight, but the pilot was able to make the necessary contact with Shannon Airport in Ireland. They landed there and Mike got to spend some time in Ireland. Then on to Burtonwood, England. Mike made two flights into Germany with BX supplies. On orders dated 23 Sep 48, he left Frankfurt, Germany and arrived back at Wright Patterson on 25

Sep 48. Mike was happy to participate, in a small way, with the Berlin Airlift. He was not there long enough to qualify for the medal that was bestowed on the participants who were there 90 days or more. Mike met several people whom he had flown cargo with in India. He purchased a cuckoo clock and managed to get it home in one piece.

Mike had been going to Huntington occasionally and now he stopped altogether. Mike and Martha became engaged to one another with the wedding set for January 8, 1949. Martha was kept busy working out the details of the wedding. She sent out announcements, purchased a gown and flowers, and worked herself into a tizzy to get the details straightened out. They were married at the base chapel at Wright Pat and held a reception at the Wishing Well. Many of their friends from Wright Pat were in attendance as were relatives from Marion, Ohio and Huntington, West Virginia.

Mike and Martha rented an apartment near the base and settled down to a routine existence with very little travel involved. Mike and Martha went fishing at some lakes on the base. Mike put in his application with the West Virginia State Police and found out that he was overweight. Martha was instinctively a good cook and kept Mike well fed. They spent some time at the club and visited with friends. Mike took his discharge in May 1949 and moved to Huntington.

Mike went on a strict diet to lose 15 lbs. by the last of June. Mike and Martha moved in with Mike, Sr. and Sadie until Mike could graduate from the academy. The folks were glad to provide temporary quarters for Mike and Martha, but all new brides like their own place and Martha was no exception. Mike passed his physical and was sworn into the State Police in June. He worked out of the South Charleston barracks until the academy started the training program.

He reported to South Charleston Company B Headquarters. While talking to the sergeant, a call came in reference to a recovery of a body near Saint Albans. Mike went with the sergeant and it was a gruesome sight. The water was warm and the victim had been in the water for over a week. The body was swollen to twice its size. Large blisters had formed on the body and it had been chewed on by turtles and other aquatic life. It was a very unsavory sight. The smell was atrocious. They zipped the body into a plastic bag and the sergeant took some notes. The sergeant then said it was his lunch time and drove down the road to a restaurant. He ordered a hamburger and asked if Mike wanted anything. Mike managed to reply with a weak, "Nothing for me, sir".

CHAPTER 6

Madison

The Class of 1949 was the first to graduate from the State Police Academy. The building wasn't completed when the class reported and the cadets helped finish it. The Academy was a nice modern facility and everyone was proud of it and glad to help with the work. Mike was the oldest man in the class and one of the few who had attended college. Graduation was not a cinch and the troopers did a lot of preparation for the various subjects. The Department hired a cook and the food was very good. Lights were out at 10:00 p.m. with all the good food and plenty of sleep, the cadets were in exceptionally good shape.

The class started in September 1949. On October 4, 1949, Mike received a call that he was the father of a baby girl. He didn't lose any time getting to Huntington to congratulate Martha. There was another trooper from Huntington at the academy. He had fire engine red hair and he answered to the name of Red.

One night the cadets were telling jokes. Red was in the shower. The lights were out. Another unit had just finished taking a cold shower and was looking for his towel. Red heard the other men laughing and stepped out to see what was going on. He was feeling his way in the dark and he put his hands out and felt the wet, clammy body of the other cadet. He let out a yell and ran out into the sleeping bay, claiming that he had felt a ghost. While Red had been in the shower, Mike set the stage for a playful prank. He told the barracks full of men that when Red came out he (Mike) would tell a joke and the men were to laugh whether it was funny or not. Mike told a dry, British joke about someone stuffing French fries in their ears. Something about the pathetic joke and the circumstances before it was told struck the men as exceedingly funny. They started laughing and couldn't stop. Red sat on his bed laughing up a storm. Finally things settled down and the barracks became quiet. All a once Red came alive and said "Hey! Casey tell that joke again, I can't figure out what I was laughing about". The barracks came alive again with hilarious laughter.

One of the men chewed tobacco. He could go the whole class

without spitting or getting sick. He could take a drink of water or a bite of food with tobacco in his mouth. One of the training officers also chewed. Nothing was ever said about the cadet chewing.

One of the troopers had a foreign-made automatic pistol, which he kept hidden in his trunk. He enjoyed getting it out and cleaning it. One night he was showing off his weapon and accidentally fired it. He swore that the weapon was on safety. He was a pretty good marksman and hit himself in the broad part of his foot. He was still having trouble walking when the class graduated.

Shortly after the academy was ready to accept students, Mike was summoned to the office to discuss an assignment. It seemed that things were getting out of hand at Point Pleasant. The private clubs were not doing a good job of screening the patrons that used the clubs. The prosecuting attorney requested an investigation. The district sergeant at Point Pleasant requested a trooper be assigned to his office for some undercover investigations. They picked Mike for the job because he had an Air Force ID card and could use it to advantage. The same weekend Mike was supposed to move on the activities, a friend of his drove to Huntington in a new convertible coupe. Mike and the friend had been stationed together in Japan. In fact, Smokey had worked for Mike in the communications section. Arrangements had already been made to meet the district sergeant at the state road garage. The sergeant gave Mike $100.00 to use in the investigation. He filled Smokey's gas tank and told Mike where he wanted them to go. They noticed a cab parked on Main St. and got into a conversation with the driver. The cab driver said his name was Ace and said that he had access to some whiskey. Mike gave him $15.00 and told him to get them a fifth of whiskey. They locked the fifth in the vehicle and went to one of the clubs and had no trouble getting in. They played the slot machines until they got a pay off. Mike got into a poker game and lost $20.00. They went to several other establishments and were successful at all of them.

The next day was Monday and Mike went to Mason County and signed warrants for various individuals. He turned over several fifths of whiskey that he had purchased. Several city and county officials were present at the clubs and were later arrested and pled-guilty on the following Monday. The district sergeant called in several troopers to raid the establishments. Mike had to testify in court not long after that and, needless to say, he was not popular around Point Pleasant. It all worked out well. Mike got to spend some time with Smokey and conduct a good investigation at the same time.

In the interim between graduation from the Academy and going to Madison Mike was assigned to Barboursville. Even though he was 29 years old, they treated him like a rookie. One trooper put some rocks inside the hubcaps of Mike's police car and when he started down the road it sounded like his car was coming apart. Mike was serious about his job and equipment assigned to him and returned to headquarters and told the men that something was wrong with his transmission or perhaps the universal joint. The troopers got in a huddle and started a big discussion about all the bad things that could happen to a police car. They talked about misuse of department equipment and about liability if the trooper was at fault. Finally, after they had milked it all they could, one of the units went out and brought the hubcaps and rocks into the barracks. Then they started guffhawing, and patting each other on the back. They cackled and laid eggs all over the barracks.

A lady called in a complaint. She stated that a black man about 70 years old was trying to relieve her of some cash. The lady had a brother in the penitentiary and the man told her he could get him pardoned for a fee. The lady gave him $50.00 and promised him another $200.00 when her brother was a free man. The lady got to thinking after the con man left and called the police. The officer on duty thought he would have some fun and he assigned Mike to the case.

Mike took it seriously. He went to the lady's house and took notes. He obtained a description of the man and learned that he was being driven by a younger man. Mike went on down the road checking for the man. The suspect stopped at another home and asked if they had relatives in the state penitentiary. The suspect offered to get the person out of the pen for $50.00. He then asked where the nearest Greyhound bus stop was. He was told Milton. Mike went to Milton and was told that the bus had just left and was headed for Hurricane. Mike rushed to Hurricane and caught the bus just as it was leaving. The suspect tried to get off the bus, but Mike placed him under arrest and took him to the Barboursville police office. The suspect was searched. He was wearing long underwear and had little cloth pouches pinned to his underwear. Inside the cloth sacks were pieces of paper with the words MONEY, MONEY, MONEY written on them. The troopers were flabbergasted. Since Mike was leaving for Madison soon, the sergeant turned the investigation over to the trooper who had received the call. It meant more work right at meal time and Mike had the last laugh.

On January 1, 1950, Mike received orders to proceed to Madison, West Virginia. Madison was a small coal mining town of about 2500 people and county seat for possibly 25,000 people. It was a pretty rough area, especially on weekends. Mike's time off ran from 5:00 p.m. on Wednesday until Thursday at 4:00 p.m. A member of the Department of Public Safety was expected to devote his entire time and attention to the business of law enforcement in the State of West Virginia.

In order to get his family with him Mike accepted substandard quarters. He later regretted it. The apartment was up over a noisy restaurant and was equipped with wall to wall cockroaches.

Mike's first full day as a trooper in Madison fell on a Saturday. Most of the court house offices were open on Saturday morning and Mike spent some time getting acquainted with other officials. He was standing on the sidewalk in front of the court house when he found himself lifted off the sidewalk and swung around like a toy. Mike looked down and saw two big wrists around his waist and he struggled to get loose. He was finally successful and found himself looking at the biggest woman that he had ever seen. He was very much disturbed and just a little angry at this undignified conduct. The corporal in charge of the post came by and told the Amazon to display her strength somewhere else. Mike walked off slightly embarrassed. They found out later that one of the deputies had put her up to the bear hug.

The first weeks in Madison were not pleasant ones. Mike soon had his first fight as a state policeman. He had been to the movies with Martha and was in civilian clothes. A miner on holiday yelled at Martha, "Come on over here, Red." Mike asked him if he had yelled at his wife. The miner replied, "You damned right I did." Instead of walking away like he had some sense, Mike placed him under arrest for intoxication. The miner had a buddy who jumped on Mike's back. Mike defended himself against the miner that did the yelling. It began to get wicked. Martha ran to the car to get Mike's weapon. Mike knocked the first miner down and his head started bleeding. He hit the second miner and knocked him down. Then he handcuffed the miner he had arrested and took him up the street to the jail. There was a big crowd of people shopping. They lined the streets to watch the excitement. Mike took the bleeding miner and booked him on a drunk charge. The jailer called the corporal and told him what had happened. The corporal took a statement from the tavern owner and told him if he couldn't run his business that they would do it for him. The next morning the miner pled guilty, paid a fine, and apologized. Mike's reputation as a no-nonsense policeman was enhanced.

The corporal and Mike went to the Seth area to investigate a complaint. It seemed that a tavern owner became intoxicated in his own establishment and was waving a loaded revolver around his head and threatening patrons. The two officers were in the general area and pulled in just below the tavern and parked the car. They entered the establishment and the owner threatened to shoot the corporal if he didn't leave. This was the wrong thing to do. Mike pulled his revolver and drew down on the owner and told him to lower his gun and lay the weapon on the counter. The owner must have sensed that he would be fired upon point blank if he persisted. The corporal stepped up to the owner and put his finger between the hammer and the firing pin. The owner capitulated and the corporal took his weapon. He didn't get mad at the corporal, but he told Mike that next week he would be pissing in the Yukon River, implying that he had enough political pull to send him to Alaska.

The state police in West Virginia have full police powers and can enforce any law on the books, even game violations. Mike once arrested a subject for wearing a Masonic ring. It was definitely a violation of the Code of West Virginia. No other lodge ring was protected to this extent. The law had been on the books for years and years and was never enforced. Mike was investigating the subject on another charge and lodging him in jail made sense. Mike had plenty of time to get a search warrant and search the subject's vehicle for evidence vital to the investigation.

There was never a dull moment. The state police office was located in one large room in the courthouse. There were always people around. Saundra was growing like a weed and developing an outgoing personality. Mike didn't get to spend much time with her. They found a nice garage apartment to live in and it had a fenced in backyard. The family they rented from lived beside them. They loved kids and helped look after Sandy.

Mike had put his Air Force career on the back burner. He had not been involved with any type of training or other activity. He joined the Reserve Unit in Huntington and drove down once a week for training.

Mike was inclined to pull a dirty trick on occasion. A prominent member of the Reserve training squadron had access to X-rated movies and every once in awhile he would show one to the squadron. The member did not know that a state policeman had joined the squadron. The chairman of the entertainment committee called Mike and told him to wear his uniform and be prepared to make an arrest. Mike did better than that, he brought along a warrant for the arrest of the accused. Mike arrived a little late and the movie had already started. Mike immediately read the charges and put the hapless airman under arrest. Mike confiscated the film and handcuffed the accused. Mike started out to the car with the distraught airman, but stopped when he observed that the man was taking it seriously.

The state police top brass were unrelenting in their effort to get Mike separated from the Air Force Reserve. One morning in July, 1951, a high ranking officer of the department engaged Mike in a conversation on the courthouse steps. The dialogue went as follows:

"Casey, you are going to have to get out of that damned outfit."

"What outfit, sir?"

"That damned Air Force Reserve outfit, that's what."

"Why must I get out of it, sir?"

"Because! Someday you will be drawing two pensions" said the senior officer.

"What's wrong with that, sir?" Mike asked.

"It just ain't right" said the senior officer.

All this time Mike was taking care of his department duties in a capable manner.

One dreary, foggy morning a tragedy occurred in the little hamlet of Emmons. The only thing there was a post office and two or three houses. It turned out that the previous afternoon three jet fighter planes landed at the Kanawha Airport and decided to RON. The next morning there was an extremely heavy fog. They took off anyway and about five or six minutes later they crashed into a hill at Emmons - one, two, three. They were apparently in formation. It was never determined why they crashed. Mike arrived on the scene and there was utter chaos. Everyone for miles around was walking into Emmons on the railroad. There wasn't much left of the planes. Mike attempted to secure the area with very little success. The Air Guard finally took over the area.

The first Christmas that Mike spent in Madison was 1950. Martha looked all over for a present for him. She finally located a unique gift, a flashlight with a blinking light on one end and a regular flashlight on the other. The blinker was extremely handy at the scene of an accident. One night Mike stopped at Uneeda to check a tavern that had been giving them problems. Inside he noticed a young man who he had recently arrested for auto theft and was on probation. Among the terms of his probation, he was not to visit taverns or any place where liquor was sold. It was obvious that he was going to cause trouble and Mike arrested him for parole violation. Mike started to put him in the police car when he broke and ran. Mike ran after him. Mike didn't want to shoot him. So! He threw the flashlight and hit him in the back. The light fell to the road and broke into several pieces. The man crossed the road and waded a small stream. When he got to the other side, he begged Mike to shoot him. Mike knew where he lived and went to his home and waited for him to come up the railroad tracks. He was handcuffed and taken to the county jail. Later he was sent to the state penitentiary for violation of probation.

Part of Mike's patrol area in Boone County was the Bear Creek community. He drove through the area about every other day and felt the pulse of the people and listened to their complaints. Most of the discussion centered around a woman about 65 years old on relief. She had eight children, four boys and four girls. She lived with a 22 year old male and spent more money on him than she did her children. One day Mike was going down the creek and fell in behind a drunk driver. The driver wouldn't pull over until he arrived at the home of the lady who was the topic of discussion. The woman and her offspring were out in the front yard. There was a creek between the road and the old woman's house. There was a walk log across the creek. The drunk driver threw a bottle over the embankment into the creek. Mike needed the bottle for evidence. He got out of his vehicle and conducted a search. He finally found the bottle and held it up for the old lady to see. Mike asked her if she had seen the suspect throw the bottle. The woman turned red in the face and shouted, "I didn't see a G-- d---- thing and I wouldn't tell you Big Hats if I did." Mike put the whiskey bottle into his trunk and the drunk into the front seat of his police car. The drunk had another white male with him who went over and set on the front porch. It turned out that this was the youth that the woman was keeping, Mike told the old woman and her live-in boy friend that he would see them later. It was sooner than he thought. That very night he was patrolling down the creek and saw a pedestrian. Mike picked him up and learned that he was the son-in-law of the old woman. Mike saw a chance to clear up the whole mess.

Mike asked the son-in-law if he had ever seen the two subjects in bed together. He said yes, that he had seen them on many occasions and that he didn't believe in exposing the children to that kind of conduct and spending money on wine that should spent on the children. He agreed to sign a warrant, charging lewd and lascivious co-habitation on each of the subjects and to testify in court if necessary. One of the elements of such a charge is that the subjects be caught in bed together.

Mike took the son-in-law before a Justice-of-the-Peace and he signed a valid warrant calling for the arrest of the two subjects. Mike took him back to the creek and told him to go in the bedroom and turn on the lights and Mike would stand at an outside window and view the scene. The son-in-law went through the unlocked front door. He must have stepped on the children for they began to mutter and mumble. He went into the bedroom and turned on the light, which was just a cord from the ceiling with a bulb on the end. Mike saw all he needed to make the arrest and place them in custody. Mike sent the son-in-law after his wife and she took over the kids. Mike took the couple to jail. The two pled guilty to the charges and were placed on probation. One of the terms of probation was that the young man stay away from the premises and not contribute to the delinquency of minor children. She was to accept supervision from her oldest daughter about the spending of money. Mike got a conviction on the drunk driver. He lost his driver's license for six months and paid a $50 fine. Bear Creek was quiet for some time to come.

One night about 10:00 p.m. the lady who owned the apartment Mike lived in called Martha and told her that a neighbor, who was a disabled miner, was drunk in his home and that he was carrying a gun and firing it in the air. Her husband was due in from work shortly and she was afraid the drunken neighbor would shoot him. Mike had jurisdiction. But! He didn't like to get too involved with town enforcement. As Mike walked up onto the porch he could see the man sitting in an easy chair. Mike walked in the room and the man reached down beside his chair. Mike jumped clear across the room and grabbed the man's arm. On the end of his arm was a long-barreled pistol. Mike could have been shot. He arrested the man for brandishing a deadly weapon and for disturbing the peace.

Producing and selling moonshine whiskey was not uncommon in Boone County. During the three years Mike was in the area, he ran across three or four stills. He never caught a moonshiner at work. They had a communications system that seldom failed. The neighbors always protected them. There was never a path to the still. The shiners always came to the still from a different direction. They generally located near a spring in isolated places. A still usually consisted of two mash barrels, a copper coil, and a boiler to heat the mash. The mash usually consisted of corn allowed to ferment. The vapors from the boiler went through the copper coil which was cooled with water. This turned the vapors into whiskey which dripped out of the bottom of the coil. Squirrels and raccoons, birds and other animals were attracted to the mash. Many times the animals would fall into the mash barrel. Once in they seldom got out. The shiners never let dead animals bother them too much. They just lifted them out and threw them over the hill. Mike sometimes carried a half stick of dynamite and would literally blow them up. Some families took pride in their endeavors and produced some good corn whiskey. The roads were so poor in West Virginia that they could never haul the corn out in a truck. So! They reduced it to a liquid state and it was easy to transport.

Payday in the mine fields was always hectic. The miners were mostly paid in cash. The state police in Madison had traditionally carried the payrolls over the mountains to Prenter.

They carried 40 to 60 thousand bucks cash in the trunk of the patrol car. The corporal said he would rather haul the cash than look for it. They were still hauling the cash when Mike left Madison.

The commander of Company "B" called for Mike to meet him at the Kanawha/Boone county line. The commander was a very capable man and knew the pitfalls of traveling through another state with a prisoner in tow. Since the state you are in has jurisdiction, the officer could be charged with manslaughter or murder if he wasted a prisoner. He explained to Mike the stigma attached to letting a prisoner escape. The commander wanted Mike to know all the facts and balance them against the prospects of a good trip and decide whether or not he wanted to go.

Mike decided to make the trip to Huntsville. The West Virginia probation and parole authorities were footing the bill. They gave Mike enough transportation vouchers to travel by bus, plane, or train. Mike flew to Dallas and obtained a room in the Shamrock Hotel. He then called the Texas Rangers and told them his mission. The next morning a ranger captain picked Mike up at his hotel and drove him to Huntsville. Mike met the prisoner and became the custodian and signed for $40.00 in cash that the State of Texas owed the prisoner for work he had done. They gave him a set of khakis. Mike and the captain had lunch at the prison and started back to Houston. The airlines at that time would not let a prisoner fly as a passenger if they were aware of it. Mike stopped at a department store near Houston and bought the prisoner a shirt and trousers and a pair of shoes.

Texas prisoners were put into jail as punishment for crimes and were not to be mollycoddled. Authorities used some harsh methods to keep them in line. On the way back, the prisoner told Mike some horror stories. It seemed they had several classes of trustees. The elite of this group were called high riders and some of them were said to be armed. It was also rumored that if a high rider killed a man while he was trying to escape he was given a pardon. Another group of trustees were called dog sergeants. They were given several months off their sentences if they prevented an escape. The prisoner did not wish to linger long in a Texas prison. The man was an escapee from the Texas prison system. He had also escaped from a West Virginia county jail after commission of a felony.

He didn't look good, but he didn't look like a prisoner. Mike put the prisoner in jail and made arrangements to fly the prisoner to Charleston. The schedule called for them to leave Houston at 3:00 p.m. and fly to New Orleans and stay all night, then fly into Charleston the next day. When they got to the airport, the prisoner balked and said he wasn't going to fly. Mike was tired and wanted to get home. Mike got the prisoner into a small room at the airport and read the riot act to him. He told the prisoner he was going to take him by train and stop off and put him in some little tank town jail each night and take several days to get home. The prisoner had been in southern jails before and knew how dangerous they were. He decided to cooperate. Mike handcuffed the prisoner to himself, put a raincoat over the handcuffs, and they got on the airplane. Mike decided to spend the night in New Orleans. He put the prisoner in a precinct jail and got a hotel room for himself. Two detectives asked him if he wanted to work with them that night and see the town. Mike was eager to get started. He saw the town of New Orleans as very few others see it. The next day he arrived home with his prisoner.

One dark and stormy night, Mike was patrolling down Route 3. It was about 9:30 at night and a basketball game between Lincoln County and Madison had just ended a few minutes earlier. A vehicle going south had a flat tire in a bad curve . A vehicle with two boys in it stopped behind the car with the flat. The boys were between the two cars getting the spare out when another vehicle came around the curve and hit their vehicle. One of the boys managed to escape, the other was pinned between the cars and suffered a severely injured leg.

There was no time to wait for an ambulance. Mike put a man with a light on each end of the curve and another spectator to direct traffic. Mike headed for the hospital as fast as possible with the victim. He called Charleston on the radio and ordered two wreckers to the scene and also informed the hospital of the condition of the victim. The hospital was small, but had a good doctor on duty. The doctor took one look at the leg as Mike and the only nurse on duty got him out of the car and onto a gurney. The doctor wasn't bashful. He ordered Mike to get the victim on the operating table and cut his pants off. The smashed leg and foot must come off immediately. There were no attempts to be fancy. The surgeon cut the leg off at the knee joint, pushed the

Trooper Casey and Governor Frank Carlson of Kansas
Governor's Conference at White Sulphur Springs

flesh up on the thigh bone exposing the knee joint and about two inches of bone. He cut the bone off and let the flesh come back over it. The doctor stated that later on the victim would thank him for leaving him a pad on the knee.

Headquarters sent an officer over to Madison. He was about ready to retire and Mike and the corporal wondered about his effectiveness. One night about 10:00 p.m. Mike was in the courthouse with the new officer when they received a call that a man wanted by the police was at a certain location at Pratt. Mike went with the new officer to Pratt. They went up onto the front porch and Mike saw the suspect lying in bed. He told the other officer to guard the front of the house. Mike would go in the backdoor. Mike went to the rear of the house and the door was unlocked. Mike walked through the house with his flashlight and went into the bedroom.

The man was in a drunken slumber and Mike hit him sharply with his open hand and the man jumped up. Mike happened to flash his light over to the window and saw the elderly officer with his 38 cal. pistol pointed straight at the suspect's head. Then Mike noticed a large knife on a small night stand. Mike is convinced to this day that if the suspect moved a muscle toward the knife, the officer would have burst a cap on him. Mike never had a better man to work with.

In 1950 the State of West Virginia hosted the Governors' Convention. It was held at White Sulphur Springs. Mike was detailed there for several days to serve as escort for Governor Frank Carlson of Kansas. Governor Carlson was chairman of the organization and he arrived several days early and stayed a few days extra to handle certain things for the next convention. Mike really lived it up. Wonderful meals. His duties were light. All he did was drive the governor in a brand new Ford wherever he wanted to go. Governor Carlson had his daughter and son-in-law with him and Mike drove them to several places of interest. Mike had a good time, but he was glad to get home to his family.

A man came into the office one evening and complained about his wife assaulting him. Mike told him that he hesitated to get into family arguments, but if he wanted to get an assault warrant it would be served. The man was of average height and build and did not appear to be a weakling. He stated that very few days went by that his wife didn't hit him with something and if she couldn't find anything else she would hit him with her fist. According to the victim, this had been going on for years.

The man obtained a warrant and Mike locked up the office and prepared to serve it. When he got to the man's home on Pond Fork, he found that they were having a fight and the wife was doing most of the fighting. When Mike produced the warrant and told her she was under arrest, she let it all out. Her son was taking a woodwork class in school and had just completed a lamp. He turned it out on a lathe and it was heavy. While Mike was trying to protect the husband, the woman hit Mike on the head with the lamp. Luckily he had his hat on and all it did was raise a big lump on his head. Mike managed to handcuff the woman and take her to jail. Then the man got into the act and wanted Mike to turn her loose. The man posted bond for her. Mike learned a big lesson. Stay away from domestic arguments! Years later Mike had to have a cyst removed from his head. He felt certain that it was caused by the lamp.

One night while patrolling the Pond Fork area, Mike arrested four drunks. It takes no great skill to arrest a drunk, but in the coal fields it makes the difference in getting a good night's sleep or no sleep at all. If drunks were left out on the town, they would invariably get involved in wrecks, assaults, rapes, thefts, etc. and Mike would get called out to investigate those crimes. This type of policing is good for the drunk. If left to his own devices, he might get run over or fall off a bridge, break into a store, or get hurt in a fight. Mike put three of the drunks in the back seat and one in the front. He neglected to search them. When they got to the jail, the jailer took a cheap owl head out of the coat pocket of one of the drunks. The pistol wasn't worth 15 cents, but it could kill if someone pulled the trigger. Most policemen get a little careless at times, but Mike learned his lesson.

A Boone County serviceman was home on leave and was out having a ball at the local tavern. There had been a lot of drinking going on and the tavern operator closed up about 2:00 a.m. The serviceman apparently had one too many and went to sleep in his car. He woke up about 2:45 a.m. and got out of his car. He was parked close to the road and a vehicle came along and hit him. He wasn't hit very hard and there was very little damage to the vehicle, an old truck. The man that hit him lived in the area, and he stopped and could see that the serviceman was dead. The victim was dragged for a few feet and

Mike, Sr. and Sadie Casey bound for New York,
Thanksgiving 1951.

had several injuries. His service cap had fallen off and for some strange reason the driver decided to throw the cap in a two hole outdoor toilet. The driver then left the scene and did not report the accident. Mike conducted a full scale investigation. For the next three night in a row he went to the scene at the same time the soldier had been killed, hoping to find a shift worker or someone who came through the area at the same time every night. It paid off. On the third night the accused came through and Mike stopped him. After a very brief conversation, the accused said he wanted to confess to the crime. He could see by the three nights Mike had spent on the scene that the authorities were not going to turn loose. He said besides that his conscience was hurting him. They recovered the hat from the two-holer and Mike placed the driver under arrest.

On Thanksgiving Day of 1951, Mike got a call from a lady in Berlin. She stated that she had reason to believe that her son was going to commit suicide. He had received a bad discharge from the Army and had mental problems. The family had Thanksgiving dinner at the mother's home. The subject took offense at something said at the table and took off for his home. He picked up a light jacket and an old shotgun. He didn't take his lunch. He didn't report for work on the evening shift. He was never seen again.

The winter of 1951 was the year of the big snow. It started snowing about 4:00 p.m. and didn't stop for a day and night. In some places the snow was three feet deep. It was impossible to hunt for the missing miner. Several months later in the spring two boys ran into what was left, which was just a few bones and some articles of clothing. It appeared the subject had found a windfall, or perhaps he knew it was there. A windfall is where a large tree uproots and leaves a sort of cave back in under the roots and provides a certain amount of protection. He had forced the shotgun into the tangled roots with the barrel sticking out. He did a good job of securing the gun because it was still where he had positioned it. He cut a small sapling with a little notch in it and very carefully whittled it down and smoothed it out. He loaded the weapon with one shell and stuck the barrel in his mouth and pushed the trigger with the notched sapling and his head just exploded. There wasn't a piece of his skull left that was bigger than a dime. There were no pellet marks on any of the skull pieces which indicated that the shotgun barrel had been fired into his mouth. The animals had eaten all of the body. A part of a skeleton was all that was left. A tee-shirt with a bone on it was found over on the next hill.

Mike was laying in bed one morning, hoping the phone wouldn't ring, when it did. A lady said she had a neighbor on a hill back of her house that had not been seen for several days. She never said why she couldn't look for herself. Mike went to check it out. It was obvious the old lady was dead when Mike looked in the window. The windows and doors were all locked and Mike decided to break a panel of glass and unlock a window. No sooner had the pane of glass been knocked out when all hell broke loose. There must have been two dozen extremely thirsty cats trying to get outside at the same time. It seemed that the old lady had a heart attack and fell back on the brass bed. Her arm probably hit the brass bed and bruised it. The cats started eating on the bruised spot and ate most of her arm and shoulder. By now it should be obvious why Mike doesn't like cats.

Martha was pregnant with her second child. Larger quarters were needed. Mike moved his family into a nice home and got ready for the new arrival. The months passed and Martha gave birth to another beautiful daughter. Mike was at the hospital when she arrived and she came out slick as a whistle.

Mike and Martha decided they needed a home of their own. Mike bought a lot in West Madison. He contracted a laid-off mine carpenter to build a house. A helper was hired. Mike had a friend who was a drilling contractor. The friend told him to order anything he needed wholesale. Another friend with a truck went to North Carolina and brought back all the lumber he needed at wholesale. Mike hired a plumber who worked for the Board of Education by day and for Mike at night. He cut a lot of corners and still came out with a beautiful house. As a crowning touch he went to a planing mill in Logan County and got enough red wood to trim the entire house. It was beautiful!

They lived in the house for three months and the department moved them to Huntington. Mike was worried about selling the house. A man came around that evening and looked at the house. He asked Mike for a price. Mike was beside himself, afraid he was asking too much. Mike said $10,750 in a loud voice. The man never blinked. He said he had to go. The next morning he called Mike and told him not to let anyone else look at the house.

Mike and Martha had just cleared more than $3,500, better than a years pay.

The transfer as Mike saw it was made because the department wanted to please Mike's father, who was by now a prominent politician. Also they wanted to get Mike out so he could not homestead. In other words, they did not believe in owning property. Many, many state policemen in the old days spent 25 years in the department and rented a home all the time. When they retired they didn't have enough money to get back where they came from.

Mike resigned from the department and intended to make a fortune in the insurance business. The insurance company sent him to Chicago by train for a two week training involving the job. Mike didn't like Chicago and was very happy to get back to the hills. Mike had a large area in East Huntington and Pea Ridge that he was responsible for. This area was called a debit and Mike traveled it about every two weeks and collected premiums and serviced insurance policies. He did a lot of walking. He didn't last a year. He was making a little money, but hated the job. Martha helped him with his bookkeeping or he wouldn't have lasted a year. Mike pulled all the strings he could to get reinstated in the state police. Mike, Sr. was running for the State Senate with a good chance of getting elected and the department let Mike re-enlist. They sent him to Chelyan, one of the least desired posts in the state.

Mike never seriously considered moving his family to Chelyan. It was noisy and crowded and the crime rate was second to none in the state. Decent housing was not available and the detachment commander did not like Reservists.

One of the troopers warned Mike to be very cautious that the detachment commander was out to get him. The fact that he had attended college and was a graduate of the Armed Forces Industrial College and Squadron Officers School made no difference. He was behind the eight ball with the policymakers.

Mike was at Chelyan for several months and never liked it. He wasn't thrilled about going to the turnpike, but it would be nice for the family. Mike would work shifts of eight hours and get to spend more time with the girls.

First class to graduate from the State Police Academy, 1949.
Bob Casey: 2nd row from top, 2nd from right.

CHAPTER 7

The Turnpike

When the West Virginia Turnpike opened in 1954, Mike was put on orders for Princeton. He was assigned to patrol the turnpike. He rented a house within 100 yards of the pike's terminus. The house was pleasant with a good furnace (Princeton was a cold place in the winter). Behind the house was acre after acre of meadow and forest land, providing the family with a serene atmosphere in which they thrived. Mike and his family spent a lot of time walking in the hills. The job, however, was dull. It consisted of running back and forth over a portion of the turnpike, but it beat peddling insurance. Usually three patrolmen were on duty each shift and they were responsible for 33 miles of a 98 mile turnpike. Mike's usual patrol area was milepost one to 33 and included all of Flat Top Mountain which produces some foul weather patterns. The troopers on the turnpike got new patrol cars each year. The vehicles generally had 70,000 to 90,000 miles on them. They were mostly General Motors products with overdrive, specially built high speed transmissions, positive traction, and would do 90 mph in second gear.

The turnpike was usually so dull that the troopers would go to any extreme to create some excitement. One night when Mike was on the mid-night shift, headquarters sent him down to make one run from milepost 33 to 66 which included Mossy interchange. Mossy was a very lonely place at night. Two vehicles an hour would be a likely average. Mike decided to go off the turnpike and chat with the tolltaker at Mossy for a few minutes. When he neared the toll booth, Mike could see that the gatekeeper was asleep and he decided to shake him up a bit. He had an oily, old rag in the back of his police vehicle. He lit it with his cigarette lighter and waved it around to get it glowing. He wrapped another piece of cloth around it and put out the flame and left it to smolder. The toll booth had a sliding door and Mike slid it open and turned on the furnace to create a draft and slid the door shut. The rag emitted a lot of smoke. The tollkeeper started coughing and waking up. The booth was filling up with smoke and the tollkeeper was desperately trying to open the door like he would at home. Mike became ashamed of himself and saw that the toll keeper wasn't going to get the sliding door open. He opened the door and led the victim out, coughing and cussing.

There were some terrible wrecks on the turnpike. One that Mike will never forget, happened just before the 8:00 a.m. shift change. A young airman, his wife, and baby were traveling from a southern air base to Columbus, Ohio. The airman was driving. He went to sleep at milepost 30. The car ran off the road and threw the young airman and baby out of the car. None were wearing seat belts. The driver of another vehicle saw the car go off the road and blew his horn at Mike. Mike knew something was amiss and he drove back to the scene. It was a terrible sight. The vehicle had gone over a steep hill and hit a rock cliff. The mother had severe face and eye injuries and was unconscious. She remained unconscious for about 20 minutes. The baby was obviously a fatality. The airman had severe cuts and bruises. The airman had picked the baby up, wrapped it in a blanket, and laid it beside the mother. When she asked for the baby, she felt around with her hands. Her eyes were so bloody she couldn't see. She put her hand on the baby's chest and knew he wasn't breathing. She passed out again and didn't come to until after the ambulance came.

A check with the tollgate at Princeton revealed that the airman's wife had been driving when they had passed through the gate. They stopped at the toll area and changed drivers. The young airman drove off. It wasn't known whether or not seat belts were in use before they entered the pike. It was learned later on that the mother lost an eye. The airman's father chartered a small plane and took them to Columbus.

One of the tollkeepers had a tremendous appetite. Occasionally, some truck driver would leave food or fruit or ice cream for the troopers. There was a refrigerator in the little building at the toll plaza. Generally, it had something in it for the troopers to eat. The troopers ate what they wanted and left the rest for the tollkeepers. The tollkeeper with the tremendous appetite would routinely check the fridge as soon as he came on duty. No matter what he found he always went to the other tollkeeper and smacked his lips and said "Premier. Good food in the fridge and some ice cream." The ice cream was for the troopers. The tollkeeper ate it all. When the troopers came on duty, they were a little let down. The next day Mike discussed the matter with the head toll clerk. They went into town and purchased two cups of ice cream, one chocolate and one strawberry. They also purchased a little box of Ex-Lax. Two squares of Ex-Lax has enough explosive power to work the average adult male. Mike and the chief tollkeeper put six squares in a spoon, and melted it. They stirred it into the chocolate ice cream. The ice cream was put into the fridge to await further developments. One of the other tollkeepers was in on the adventure. He was told to be sure and ask for the strawberry ice cream.

The men brought back a box of cookies to be split between the troopers and a bottle of Coke. Mike was carrying the explosives. When the hungry tollkeeper arrived, he went straight to the fridge. He saw the ice cream and cookies and he started smacking his lips He exclaimed, "Premier! Ice cream!" The tollkeeper had been stationed in Germany after the war and the only word he learned was premier. When he was excited or liked something, he always reverted to his extensive knowledge of German and said "Premier! Premier!" He turned to his buddy and asked him which one he wanted. The other tollkeeper replied, "Strawberry." The ever hungry tollkeeper answered him, "Premier! Premier, I love chocolate!"

After they had been at work about one hour, the ever-hungry tollkeeper went to the fridge and got the ice cream. He took it to the toll booth. He said to the other tollkeeper in the booth, "Here's your strawberry. Them damn troopers are getting too fat anyway!" They ate their ice cream and the ever-hungry toll taker apparently licked his cup clean.

Here is the sequence of events as related by the second toll taker. After about 20 minutes, the ever-hungry tollkeeper bent over and grunted. Almost immediately he said. "Lordy, Lordy! I believe I'm catching the appendicitis." And then the tollkeeper doubled up and headed for the toll house. He was gone about 15 minutes and came to the toll booth pale as a sheet and said that if he didn't get better soon he was going home. He went back to the toilet for another session and gradually began to feel better. He never caught on to what had happened. Even though all the people working on the Turnpike were told and most of the truck drivers who routinely traveled on it knew, it was months before the tollkeeper found out. Later, one of the truck drivers was talking to the ever-hungry toll taker and asked him what happened to the worker they fed the Ex-Lax to. The ever-hungry tollkeeper knew what the score was then and was always careful to check what he ate. The hungry tollkeeper suspected the right people were responsible for the dirty tricks. But! He never found out for sure.

Things were not always peaches and cream in Mike's little world. As mentioned in some of the pages about flying the Hump, it was difficult to have bowel movements under extreme conditions. Mike had delayed the function on many occasions to get back to decent facilities. This put an unnatural strain on the lower intestines and eventually caused hemorrhoids. Riding in the police car and not getting the proper exercise didn't help. It was getting painful to defecate and Mike went to the veterans' hospital. The medics agreed that long intervals without relief could have caused the problem. They sent Mike to Roanoke, Virginia for his operation. The procedure was a success. Mike sat on a rubber cushion for several weeks after the operation and was never bothered with the problem again. This made five times Mike had been hospitalized.

Mike was doing well with his Air Force Reserve activities. He received orders to report to Maxwell AFB for Command and Staff Orientation Course. The Command and Staff School was a real bonanza. Most Reserve officers had to take the course by correspondence and did not have the advantage of associating with their peers. There was a certain amount of prestige in completing the course in residence.

Mike was appointed flight commander of the two flights in Princeton. They met once a week over the fire station in Princeton. Mike was told in no uncertain terms that if he wanted to do well in the Department of Public Safety he should give up his Reserve activities.

The shift work was a bargain in one way. It enabled Mike to spend more time with his family. Mike loved his wife and daughters and vowed never to do anything that would make them ashamed of him. They sometimes took long walks on the property where they lived. They would pick blackberries in the summer and gather hickory nuts in the fall. They always were affiliated with the church wherever they were stationed.

One summer day in Princeton, they went for a long walk in back of the house. The cows had been there first and had evidently eaten lots of grass. There were little brown mounds of cow manure all over the place and it was fresh! Mike believed in Murphy's Law. Anything that could go wrong, would, and at the worst possible time. Before he entered the cow pasture with his daughters, he painstakingly explained how they should avoid stepping onto these little brown mounds. The girls, who were about five and three years old respectively, listened intently, and nodded that they understood. Mike first lifted Sandy over the fence and she waited. He then lifted little Martha, the youngest, over the fence. He then climbed over the fence himself. He no sooner climbed over the fence and turned to take the girls by the hand, than little Martha stepped into an especially fresh brown mound. Little Martha had never heard of Murphy's Law, but she understood it after stepping into the cow dung. She let out a wail that could be heard for a mile. Mike lifted her foot out of the smelly goo and her shoe remained imbedded in the cow dung. Mike sat her down so he could concentrate on getting the shoe out of the mound. She then inadvertently stepped into it again with her stocking foot. This time the manure oozed over the sock. The poor child almost lost her balance and in her attempt to right herself, she stepped back into the shoe Mike had attempted to clean. This time she managed to generously fill the inside of the shoe with the brown goo. With a sense of defeat, Mike decided to discontinue the hike and headed home to get cleaned up after the ill-fated nature walk.

Sometimes when they went on hikes Mike would take a canteen of water. The girls were enjoying drinking water from the canteen. It was very hot and the canteen was getting a lot of use. The water ran out. There was nothing else to drink. Mike kept on walking and talked about how thirsty he was and how good a nice glass of lemonade would taste. Sometimes the powers of suggestion are overwhelming. Both the girls said they needed to have a drink. They were dying of thirst and Mike kept pouring it on. "If we were home right now, I could drink a pint of ice water." said Mike. The girls said that they could drink a whole barrel full and started walking faster to get home quicker and have access to all the water they wanted. The faster they walked, the more thirsty they became. They arrived home just in time to survive the "intense heat". "Martha" was not pleased with Mike's antics. These antics took place at least two or three times a year. Mike could have them dying of thirst with just a few words.

A very amusing event took place at the Bluestone Restaurant one night about 8:00 p.m. While Mike was relaxing and holding a conversation with the toll clerk, a family came through the gate and parked near the police car. The two boys in the back seat had a baby alligator they had brought from Florida. What to do with it was the question? Mike didn't know the solution to the problem, but saw a way to have some excitement on an otherwise dull evening. After the vicious five inch alligator had headed on north, Mike went to the phone and called the manager of the Blue Stone Restaurant, who happened to be working late that night.

"Good evening, Mable. How are things at the hash house?"

"Pretty good. How is it with you?" she replied.

"Well, a family just came through the tollgate and said that they had stopped at your place of business and had a problem." Mike stated.

"What's the problem?" the manager said a bit alarmed.

"They had two sons that brought a pet back with them."

"What's that got to do with me?"

"The pet was an alligator."

"I say again what's that got to do with me."

"The alligator escaped when they took the lid off the box in your dining facility."

"Oh! My God! How big was it?"

"About 18 inches."

"Will it bite?"

"Has a rabbit got ears? Yes! If you happen to catch it, open a box of corn flakes. The gator loves corn flakes."

Mike could hear them moving marble machines, and music boxes, and going besirk. Several truck drivers were there. They didn't know what was going on exactly, but they very effectively helped keep things in an uproar.

Fog was always a potential problem on the turnpike, especially at night. Mike decided to take advantage of it. He was at milepost 22 and bored stiff. He decided to create a diversion. There was a new radio dispatcher on duty at the Charleston end of the Turnpike. The radio operator was supposed to provide information to the troopers, log all calls, and was supposed to be in control, and provide any service required by the troopers. Mike called him knowing he had only been on the job about a week. The new radio operator was no match for three troopers in a playful mood. Mike's unit number was 295.

"Unit 295 to headquarters."

"This is headquarters, go ahead 295." said the radio operator.

"The fog's beginning to move in. We may be in for a rough night." said Mike in a serious, official tone. Actually, Flat Top Mountain was clear as a bell.

"Keep us posted Unit 295. Over and out." came the reply.

"Roger, over and out."

Five minutes later, Mike called, "Unit 295 to headquarters. The visibility is getting rather poor. We may need the fog grinder at any time."

The operator asked, "Unit 295, this is headquarters. What's the location of the fog grinder?"

"Unit 295 to headquarters. The last I heard it was somewhere around Mossy interchange." stated Mike.

The radio operator called Mossy and queried, "Headquarters to Mossy. Is the fog grinder at your location?"

"Negative, I haven't seen it for several days." was the answer from Mossy.

All the stations with radios were playing their parts pretty well with no coaching.

"Unit 266 to headquarters. Sometimes it is left at the tunnel." a voice volunteered.

"Tunnel to headquarters. It's not here I heard that one of the sky hooks was malfunctioning." added another helpful soul.

"Unit 295 to headquarters. I heard they were having trouble with the sky hooks and it could be at the Beckley maintenance area for repairs." interjected Mike. By this time the headquarters radio operator was really trying to get in the act and help the troopers. Mike figured it was time to end the charade.

"Unit 295 to headquarters. The fog is beginning to clear up a bit, maybe we can get along without it. Make a note on your log that the fog grinder could not be located."

Mike was on patrol one night during the winter of 1955. It was the midnight shift and a very heavy snow was falling and the roads were like polished glass. Mike was carefully driving down the hill when he saw a vehicle stopped in the middle of the road. A man was lying on the road about 30 feet down the hill from the car. Another vehicle was in the ditch with the dome light on. The headlights were out, making it awfully difficult for Mike to straightened things out. The man laying in the road was dead. The man in the vehicle was dead drunk. The car with the dome light on had two children and their mother inside. The mother was entertaining them to keep them from getting alarmed. It seemed that the following events took place

The victim and his family had been laid off from a factory in Cleveland, Ohio. They were going home and had all their worldly goods in a rack on top of the car. He skidded into a ditch. The rocks and trees tore the rack all to pieces and the family goods scattered to the four winds. The man was poor and felt he could not afford the loss of his belongings. He put a pair of coveralls on over his clothing and went out to recover his pitiful possessions. It just was not meant to be. A vehicle, being operated by a drunken driver, hit him broadside right at the door on the driver's side. The man was killed instantly and left a big indentation in the side of the car. The mother stayed in the vehicle and probably was pretending everything was alright when she knew she had not seen her husband for close to ten minutes. Mike drove his police car down and blocked the view of the automobile and the victim lying in the road. Some truck drivers had stopped and put out flares. Bless them!! They seemed to always know what to do in an emergency and usually did it. Mike had called an ambulance earlier and it had not arrived. He got a minister and his wife to transport the mother and her two fatherless children to a funeral home in Princeton. Mike knew the people who ran the business would treat them with sympathy and understanding. They would help the mother and children get in touch with their friends and relatives at home.

One night just after Mike had reported for the midnight shift. He fell in behind a speeding Buick and was not gaining any ground with his police car wide open. Mike knew what to expect on the turnpike and he knew there could be some ground fog near Bluestone. Mike kept right on his tail and all at once all he could see were tail lights as the Buick ran into the fog. Mike got him stopped and it was a young man with his wife and year old baby. The man said he and his wife started an argument and he got mad and opened the car up as fast as it would go. The young woman had a driver's license and Mike put her husband in the police car. He then had her follow him to Bluestone Service Area in the Buick. It was rather cold and Mike had the service man put the Buick inside on the grease rack where it was warm. The waitresses were brought into the act and said they would see that the baby got warm milk. Mike put the driver in jail for reckless driving. The man's wife was very grateful for the way she had been treated and wrote a letter to Turnpike Headquarters to that effect. They left the next morning with the young lady driving and the young man seemed to be sincere when he said he had learned a lesson.

CHAPTER 8

Philippi

The word soon got to headquarters that Mike was commander of two Reserve Training Units at Princeton. The city let the Air Force use a room over the fire station at no expense to the government. The department decided in their wisdom to transfer Mike out of the area, far enough that he could not participate. Mike offered to go to Welch so he could stay near the area and still command his two flights. Welch was an undesirable post. The department transferred Mike to Philippi, over 200 miles away. He left with a good recommendation from the squadron commander, but no longer had the advantage of being a flight commander. He was transferred to a Clarksburg unit in August 1957. Mike moved his family as soon as possible. He located a nice house just outside of town. It was just up the hill from the American Legion Hall. The Legion Post was a very good establishment, well run, and not very noisy. Mike made three arrests there the very first night he lived in the house. It never happened again all the time Mike was there. Three drunks were fighting in the parking lot. One of them a fairly prominent citizen. The word got around that Mike was not to be trifled with. A sergeant and two troopers were stationed at Philippi to take care of law enforcement in Barbour County.

The town of Philippi survived on DPA checks and Alderson Broadus College and some mining interests. The coal business was affected by numerous disputes which many times turned into strikes. The miners struck up near Brownton. Mike went up to the picket line alone most mornings. It was a tense situation. Every morning the three Mannard brothers went through the picket line in their Jeep. Every morning the pickets surrounded their vehicle and heckled them. Sometimes as Mike drove up the asphalt road, he looked in his rear view mirror and saw little chips of blacktop buzzing around. They were using their deer rifles to intimidate and harass the officer. They were not trying to hit the police car, but it was harassment that could not be prevented. This went on for several months.

Philippi was a decent place for children to live, but had poor facilities for schools. Sandy and Martha both attended schools with outside johns and pot belly stoves. Mike was concerned that the girls would get a substandard education, but they seemed to thrive on it. They were as bright as new pennies. They had a good place to play and there were some children their ages on the dead end street where they lived. The girls recruited a marching band and fussed over who got to be the majorette.

Mike started an investigation in Barbour County that led to Cabell County. Mike asked for and was given the responsibility for the investigation. It seems a Barbour County resident developed a taste for good beef. He like it so well he didn't just get a roast. He took the entire cow. In fact, he took two or three over a short period of time. Philippi sounded like a western town with people talking about cattle rustling. The farmers were mad enough to hang someone. Some people even branded their cattle.

Mike conducted an investigation by phone and radio and was certain the beef went to the Huntington area and he had a suspect. He apprehended the suspect and headed for Huntington to try to get identification from some of the butchers. Just before they arrived in Huntington, the suspect confessed. He told Mike where the butcher shop was and Mike got a make on the suspect. Mike took statements from both the butcher and the suspect. Mike wanted to go by and see his mother while he was in the vicinity. He drove out to Sadie's house and handcuffed the prisoner to the steering wheel. Sadie didn't like it at all. "That's some poor mother's son! Bring him into the house!", said Sadie. Sadie was right about some mother's son. She fixed the prisoner some lemonade and worried about him all the time he was there. Mike got a conviction on a felony charge and the rustler went to the penitentiary at Moundsville. Needless to say, Mike became an instant favorite of the cattle farmers.

The county could not afford a full time jailer. When an officer made an arrest he put the prisoner in jail himself and waited until later to book him. The jail was old and obsolete. Business was good as there were always several people incarcerated. The trout fishing was excellent and Mike spent some time fishing. He hunted a little bit, but didn't enjoy killing small game.

Mike had his first tour of active duty 22 Nov 57 at Clinton County AFB, Ohio. It was not a fun tour. The weather was horrible, very cold. The people at the base resented the time spent with Reservists. Mike mostly laid around an overheated

Bachelor Officers' Quarters (BOQ). He was glad to get home.

The winters in Philippi were very severe. Sometimes the snowdrifts were higher than a car. The State Police office was located in the courthouse. During bad weather a lot of time was spent in the office. Mike was promoted to captain in the Air Force Reserves on 27 May 58.

While at Philippi, Mike attended Air Force Reserve meetings once a week at Clarksburg. His next tour of active duty was 10 Aug 58 at Middletown AFB, Pennsylvania, near the Hershey chocolate plant. Mike took his family with him and arranged for quarters in the guest house. The family toured the Hershey plant and the girls really loved it, especially the free chocolates passed out after the tour. Mike had a very interesting and productive tour. He helped with an investigation conducted by a former Pennsylvania state policeman. The investigation involved an individual who was taking nude pictures of some of the women at the base. He would advertise in the local paper for women who were interested in a modeling career and ask them to meet him in a hotel room to be photographed.

Mike's family really enjoyed the tour. They spent a lot of time at the Officers' Club pool. Mike took them on long drives through the Amish country. It was quite a contrast to the coal fields. Mike made some contacts that would help him get back to the same spot on a later tour. Mike completed the Squadron Officer School 24 Feb 59. He made progress during the bad weather and with his wife's help completed the course with a very respectable grade.

Mike was sent to Texas 2 Mar 59 to attend the Provost Marshalls Course. This course really motivated him. He participated in all class activities and made a good impression on the faculty and students. They nicknamed him the "Sheriff".

All this time Mike had been using his leave to get time off for training. He went to Charleston when the Legislature was in session and Mike, Sr. was a highly respected member of the Legislature. He had several terms under his belt and was Dean of the House about this time. Mike, Sr. introduced his son to the Chairman of the Military Affairs Committee, who was a Reserve Lieutenant Colonel in his own right. Labon White listened with a sympathetic attitude and was shocked at how the State Police circumvented state law. Delegate White inserted just two or three lines to a bill that read "The Department of Public Safety will upon request by a member grant up to 30 days leave for active duty training." This was hard for the Department to take. They had to comply since they were named specifically in the bill. They called it the Casey Bill. A great many of the men in Mike's class of '49 had already been promoted, some several times. One of Mike's buddies in the class was already a lieutenant which meant at least three promotion cycles.

A lengthy strike was still going on at Brownton. The strikers had become more aggressive and more violent all the time. They were destroying property and intimidating so-called scabs who wanted to work.

One night Mike had been to a Reserve meeting in Clarksburg. He took his family with him and was returning home. As they neared their home, they noticed a vehicle along the road that just a few minutes before had been blown up with dynamite. Mike's daughters had to go right by the damaged vehicle on the way to school. Mike was enraged that it happened near his home and could have injured his daughters. He put on his State Police uniform and started an investigation. He found out where the dynamite came from and who brought it into the county. He eventually made six arrests and convicted them all on felony charges. All six went to the penitentiary. This did not endear Mike with certain power groups in the county. A new sergeant was assigned to the post. He didn't like these college boys who were commissioned officers in the service. Sometimes he would get Mike's mail which was addressed in care of the State Police, Philippi. Some of the mail had Mike's rank in the address and the sergeant would ridicule him by saluting and saying "Here's your mail, Captain. You may be a captain in that outfit, but you are only a trooper in this one. Take this mail and put it on my desk, trooper." Needless to say, Mike resented this childish display.

Mike had received several complaints about a policeman trying motorists for traffic violations. He resided in a small unincorporated town and apparently worked 24 hours a day. Mike obtained statements from two individuals regarding fines they had paid on the road. Mike spent a lot of time on the investigation and learned some other things regarding the subject who called himself, chief. One night he was in the area and saw the chief's vehicle parked at a tavern. Mike stopped and entered the establishment and the chief was sitting at a table drinking beer. He was in uniform. Mike sat down and started talking to him and he became belligerent. Mike arrested him for being drunk in public. The chief was taken to the county jail. The next morning he pled nolo contendere (I do not contend) and paid a small fine. When the officer in charge heard about it, he sent for Mike and really put on a display. "We do not arrest other policemen" he roared. "You are trying to attract attention, don't ever let that happen again." He was livid. While Mike was in the office, two highly respected citizens came and stated how pleased they were to see this activity stopped.

One afternoon Mike got a call from an old man out near Audra State Park. The old man stated that his son had just committed suicide with a 22 cal. rifle. Mike went out to investigate. He found the old man back at the smokehouse. His son's body was lying nearby. The victim had a small hole in his forehead. The father was busy washing blood off the sides of bacon he had salted down in a wooden box. It seems that the victim went out to the smokehouse and shot himself in the head. He fell back over the boxes that had been used to salt and put some ashes over the bacon. The blood was red arterial blood and looked gruesome on the bacon.

Mike, Sr. was Chairman of the Institutions Committee in the House of Delegates. As such, he had to make periodic inspections of the penitentiary. One Sunday evening Mike, Sr. called and said he had to be at the medium security prison with three members of his committee the next morning. He requested a trooper be assigned to them as an escort and suggested that Trooper Casey be assigned to his committee for a day. The next morning Mike, Jr. arose early and drove to Huttonsville Medium Security Prison. The prison was well managed and the inspection went well. They traveled to Elkins in the State Police vehicle. The committee was reimbursed for their meals. They went to a nice restaurant for lunch and after lunch one of the members bought a couple of cigars and said,

"Hey! Mike can I put these on my expense account?"

"Are you going to eat them?" asked Mike, Sr. in a gruff voice. Not another word was said regarding the cigars.

A little town near Philippi was having trouble with peeping Toms. The chief of police was at his wit's end trying to solve this major crime wave. He told Mike that the city was going to fire him if he didn't make an arrest soon. The chief was about

70 years old and didn't quite have enough time to retire on Social Security. Mike liked the honest, hard-working old man and decided to spend some time helping him out. After three or four weeks Mike developed a suspect. Partly from an informant, partly by intuition, and a great deal of luck. Mike discussed his progress with the chief. The chief had known the suspect, a youth in his 20s, for several years. The chief was told to watch the suspect and if he ever caught him breaking the law to call Elkins on the telephone and have the radio operator get in touch with Mike. One dark night, Mike received a call from Elkins to get in touch with the chief. Mike was in the Philippi area not too far away and he went to meet the chief. The chief parked his car on the edge of town and was there when Mike came driving in. The chief said that the suspect was in the tavern at the bus station and was about half drunk and dancing a jig for the public. Mike arrested him on an intoxication violation and took him outside.

It was only a misdemeanor arrest and Mike neglected to handcuff him. Once outside Mike started a conversation with the chief and all at once the accused ran. He had on a pair of moccasins and he covered the ground pretty well. Mike took off right behind him and the accused turned down an alley and ran across a backyard. He had the advantage of knowing every clothesline and fence in the little town. He had been peeping for years. He ran through a yard that had a row of hedges that had been cut off about a foot from the ground. He knew the hedges were there and jumped them into the next yard. Mike didn't see them and went tumbling through the air and landed right on top of the suspect. The suspect started cussing and said one of his moccasins came off and he stopped to get it or Mike wouldn't have caught him. Mike took him over to the office and charged him with intoxication and trespassing. After about an hour session the suspect confessed and said he had been peeping into other people's homes since he was about four years old.

The chief kept his job and the suspect was fined and released. Mike's knee was never the same again. Fluid collected at the knee and had to be removed every few months. Luckily, it was investigated at the time and was considered a line of duty injury. Mike continued working even though his knee bothered him most of the time. Cortisone shots about every four or five months seemed to help it.

One morning Mike received a call from an old man near the Tucker County line. He had taken his wife to the hospital the day before. He didn't believe in banks and had an old water bucket he put all his money into and hid the bucket before he left home. He had a binful of shucked corn in the barn. He hid the bucket in the corn. He had a little over $8,000 in the bucket, including change and a few large bills. When he got back from the hospital, he took the money in the house and put it in a chair placing a large pillow over it. The next morning he called Mike and told him someone had stolen the money. He searched the house two or three times, but never looked in the chair. Mike located it immediately on entering the house. Mike took the old man to Belington and fixed him up with bank account. It was the first and only one he ever had. Part of the money was old silver coins. Some of it dated from his father and was definitely collectors' items.

Mike was transferred from Philippi to Beckley in June 1961 and was assigned to the turnpike again. This meant long hours commuting to and from Clarksburg for Air Force Reserve meetings. Mike finished the Command and Staff College just before he left Philippi. This made him eligible for promotion to major.

CHAPTER 9

Beckley

The family rented a house in a coal mining town called Skelton just outside of Beckley. Mike had located a huge Victorian farmhouse. The house was three stories high and had a basement. The house was built especially for the mine superintendent and was a very pretty place. It had a stoker coal furnace that took care of itself and provided excellent warmth. Coal was cheap at the mines and the low cost heat was a real bargain. There was a huge gob pile back of the house that consisted of coal slate and other mine debris. There were fossils by the dozens in the gob pile and searching for fossils was a favorite pastime.

New River was not far away and was one of the best fishing spots in the state. Mike did much fishing that summer. Mike was assigned to a Beckley Reserve Unit and attended the training sessions once a week.

Mike located a house on Harper Road and moved the family in. They acquired a cat that had a lot of character and was intelligent, also. The cat liked French fries almost as much as he liked getting in trouble. He got his water from the kitchen sink. One morning the girls were getting ready for school and discovered that their precious cat had little enough sense to climb a tree when freezing rain was falling. The rain covered all the limbs with ice and the cat couldn't move. It just sat there howling and Mike had to figure out a way to get it down. Mike with his bad knee had to be careful in getting it down. Finally, with the aid of a ladder, he made a dramatic rescue. Mike came out smelling like a rose on that one. He was a big hero to the girls and he liked it. One morning a car hit the cat and killed it. The cat was sorely missed by all, including Mike.

CHAPTER 10

Ripley

In the Spring of '62, Mike was transferred to Huntington. He stayed there about a month and was promoted to corporal. He was sent to Ripley as detachment commander. Mike located a nice home in about ten days and moved the family to Ripley. This was the beginning of their most pleasant time in the state. The date of transfer was Spring of '62.

Mike was ordered to attend the Provost Marshall course at Lackland AFB, Texas. He had been there once before and liked the idea of going back. Upon his return Mike found his orders for promotion to major had arrived, also an article had been sent to the local newspaper. It was a very nice article, but a trooper assigned to Ripley thought he could improve on it. He had one copy of the paper printed with the caption "THE FRUIT SALAD KID" under the picture. Martha and the girls were aware of the situation and they gave the copy to Mike. Mike hit the ceiling. He thought all the papers were printed that way. He grabbed the paper and headed for the editor's office. He was a personal friend of the editor. Before Mike could make a fool of himself the editor told him what had happened. Mike still has the copy of the paper with the humorous caption. Mike had earned a chestful of medals over the years and wore them with pride. He did not care to see them referred to as fruit salad.

There were two troopers at Ripley. Both were big and rough. Mike thought they were going to tear the building down when at lunch time they both tried to get through the same door at the same time. One trooper was transferred about two weeks later, leaving it a two-man post. The trooper who stayed with Mike was a very intelligent person, eager to work and enforce the law. One morning he came in and advised Mike that he was going to the South Charleston Headquarters and get his car cleaned up and get some supplies. Mike told him to help clean the office up and empty the trash, then he could go to South Charleston. He said, "By damn, I didn't join the department to be a janitor." Mike had about as much as he could take, he replied, "What the hell do you think I enlisted for? Certainly, not to provide valet service to another trooper." He took out the trash and helped sweep a little bit and went on his way. They got to be good friends and worked well with one another.

One night they were patrolling east of Ripley and came upon a driver who was very erratic. They stopped him and found he was drunk. His wife was in the front seat and another couple was in the back seat. Mike got the driver out of the car and told him he was under arrest for DWI. The driver started cussing. Apparently, he was the only one that was drunk. The trooper didn't like the tirade that was going on and he edged around Mike so he could hit the driver. Mike moved right with him and the trooper couldn't hit him without hitting Mike first. The trooper put the lady in the back seat and took the drunk driver to jail. When they left the trooper said, "By damn, I was just about to hit you." Mike told him to let his conscience be his guide. The next morning when the driver pleaded guilty, he told about the silver plate in his head. The trooper, not knowing about the injury, could have killed him.

One evening the jailer invited the trooper and Mike in for pizza. They had served some revocation orders and the trooper wanted to stop at the drugstore. He got some Maalox and started popping them in his mouth. He said he was getting ready ahead of time. Whenever he was going to eat spicy foods or anything that might upset his stomach, he started preparing ahead of time. After he had partaken of the forbidden fruit, he would pop the Maalox again or perhaps take an Alka Seltzer. The trooper stuffed himself on pizza and was miserable the rest of the evening. If they happened to make an arrest later, he would eat another piece of pizza as they put the prisoner in jail.

There was a man in Ripley, who didn't sleep at night. His name was Gay Winters. He lived from hand to mouth, but was a law abiding citizen. Sometimes on cold winter nights, he would sleep at one of the many churches in town. Gay and Mike respected one another and Gay helped Mike with critical information more than once.

There were several slot machines operating in the Ravenswood area. It looked like an organization from Mason County was trying to move into Jackson County. Mike was successful in breaking up this attempt by a series of raids and the destruction of the slot machines.

The county was blessed with a young aggressive prosecuting attorney, who believed in enforcing the law. He had energy to spare and often visited the crime scenes with the police. He had been a yeoman with the Navy and could pound a typewriter like nobody's business.

During the three years Mike spent in Ripley, he never lost a case in Circuit Court. Both Mike and the trooper gained the respect of the law abiding citizens of the county.

One of the first things Mike got involved with was termite control swindlers. They had taken several thousands of dollars from Jackson County residents in the past few years. The swindlers would locate an elderly couple with an old home and start to work on them. The swindlers usually visited the home and had some wood worms or other insects with them. They would crawl around under the house, then show the residents the worms and tell them their house was going to fall down around them. Some of the old folks were very gullible and would sign contracts with the crooks. The swindlers would spread oil round the house and charge the residents five or six hundred dollars for doing nothing. Most people hate to admit they are gullible and have been cheated. They simply do not report the incident to the police. Mike worked diligently on the case and sent several people to jail.

Jackson County was great groundhog country. So many of them existed that they were considered a nuisance. Most people don't realize that they make wonderful pets. Mike caught a little fellow walking up a country road one afternoon and took him home. He took over the house by evening. The girls fed him with an eyedropper at first. Later they learned that they could give him a doll baby bottle full of milk and he would sit on his hind legs and drink the milk by holding the doll baby bottle in his front paws. Nothing was ever the same after groundhog day. He established squatters rights to one of Mike's desk drawers and ate part of his stamp collection for breakfast. The whistle pig was about two months old when Mike brought him home and he grew at a rapid rate. He never laid gum drops all over the house like a cat or dog. He was a great pet until he became about ten months old and mating season came around. They finally gave him to the game warden, who released him over on the river.

Mike was in town on some business, and an informant told him that a suspect in a forgery case was in the pool room. Mike went in to check him out and noticed several people betting on a pool game. Two of them were prominent citizens. There was

Disaster control activities at Charleston Airport in 1963.
Casey third from right carrying stretcher.

money on the table. The participants made no effort to cover up the activity. Mike looked around the room and noticed several people whom he had arrested for various reasons. Mike felt he could not overlook the obvious violation. He placed them all under arrest and marched them down the street to the Justice of the Peace's office and charged them with gambling. All five pleaded nolo contendere (I do not contend) and paid a small fine. Rumors about a transfer never materialized.

Hubert Humphrey was a very influential politician and was making a bid for the presidency. He mounted a fairly active campaign which carried him into West Virginia. Mike received a call from Company "B" Headquarters advising that the feds were looking for a police officer to work with the security people. Hubert was coming in around 10:00 a.m. and would leave around 10:00 p.m. Mike was given the assignment. It was a plain clothes assignment and Mike put on his best bib and tucker, armed himself with a shoulder holster and a snub-nosed 32 cal. revolver, and headed for Charleston. Since Mike was well-acquainted in the area, he made himself useful to the security effort and had occasion to come in contact with the candidate several times. Before an appearance on TV, a couple of make-up people took over and used a powder puff to finish the job. Needless to say, Mike was impressed with Mr. Humphrey and voted for him even if he did use make-up.

Mike's daughters, Martha and Sandy, began to participate in local youth organizations, such as Rainbow Girls, 4-H Club, and others. The Rainbow Girls were involved in some sort of marching activities. The trooper was the drill sergeant. He assembled the team behind the school each afternoon and gave them close order drill. The girls loved it and so did the trooper.

Annual county-wide 4-H Camp was held at Cedar Lakes. The girls looked forward to spending the week there. Mike and Martha spent the week with them, helping with the mechanics of running the camp. It was a week of fun for the entire family.

Mike was assigned to the 9225th Air Force Reserve Recovery Squadron in South Charleston. He served as Security and Law Enforcement officer for the squadron. He had command of a ten man flight, which included one Master Sergeant and two Staff Sergeants. They met one weekend a month. Mike's participation in the Reserves was still not appreciated by the policymakers in the Department. It took twelve years to get a promotion to

Casey captures escaped convict.

corporal in the State Police because of Mike's dedication to Reserve duty.

The assignment to Ripley turned out well for the family. Mike accepted invitations to speak to local groups. The girls were getting along well in school and were healthy and happy. Martha was on the ladies bowling team and was winning trophies.

Mike's knee was bothering him quite a bit now and he decided to have an operation. He was in a Charleston hospital for almost a month. Martha and the girls visited him about every other day. It was a very painful operation and recovery was slow and tedious. Mike had to walk on crutches and then a cane. He had to go to Charleston for therapy every time he turned around. The operation left an awful scar.

One of the first people to visit Mike after his arrival back home was Gay Winters. Gay was cleaned up and had made his seasonal change of clothing. He had on the most atrocious tie in the world. Gay told one of Mike's friends at the courthouse, "If the corporal don't get back soon, they may as well close her down."

One afternoon Mike was sitting in the office. The rest of the officers were out looking for an escaped convict. The photographer was sitting in the office. The phone rang and a lady told Mike she had been driving along an isolated road and saw the meanest man in the world running along the road. Mike knew the area well and had been out there the day before to look at a drilling rig one of his buddies owned. Mike went out and moved in behind a pile of brush and sat down to see what was happening. Mike didn't have a police car and the photographer furnished the transportation. The photographer stayed out on the road. In a little while, Mike heard someone running through the brush and the runner did look like the meanest man in the world. Mike still had his knee wrapped with an elastic bandage and had to be careful. He pulled his weapon and had already decided where he was going to hit the man if he kept on coming closer. You guessed it, the right kneecap. Mike marched the man out to the road and the photographer shot a very dramatic picture and helped Mike handcuff him.

Being a policeman in a farming community is a lot different than enforcing the law in the coal fields. In an agricultural area, a policeman has time to spend getting acquainted with the citizens. Miners are a different breed due mostly to occupational hazards, which tend to make them more violent. They usually lived in substandard homes and pay outlandishly high prices at the company store. It is an entirely different atmosphere.

One day the trooper and Mike were in the office giving drivers' exams. They received a phone call from the two Jackson County deputies. The deputies had been advised by an informant that a fugitive from justice was at his home near Staats Mill. They needed assistance in arresting him. The deputies said he was armed. The trooper got so impatient he was almost dancing. He grabbed a shotgun and wanted to leave right then. Mike settled him down and told him to call and advise headquarters what had occurred. Mike grabbed some tear gas canisters and a rifle and off they went. The trooper was driving. The shotgun was between them on the driver's seat. Mike discovered that they had neglected to get shells for the rifle. His home was on the way and he advised the trooper to stop at the house for some bullets. Mike went into the washroom where the shells were. As he came out he heard a loud noise. The trooper had the shotgun in his hands and was pale as a sheet. It seems he couldn't wait to get to the scene to load his weapon. It went off accidentally and shot a hole in the top of the door and the edge of the roof of the patrol car.

The officers went to the fugitive's home. Mike had the rifle and the tear gas canister. The trooper had the shotgun and ammunition. They hedge-hopped from tree to tree covering one another all the way. When they got to the house Mike gave instructions for anyone inside to come out with their hands up. No one answered and Mike threw the canister through a window. About this time the fugitive threw his hands in the air and surrendered. He had gone to the spring to get a drink of water. He was armed with a shotgun. The troopers put the fugitive in jail and went ahead with their operator's exams. The hardest part was yet to come, explaining the hole in the door. Mike decided to be mostly honest about it and it worked. The trooper didn't have to pay anything for repairs.

Mike had an occasion to arrest a minister while in Jackson County. He did not endear himself to anyone by his actions. There were some who wanted Mike transferred, but it never happened. One of the arrests involved a tragic circumstance. A young minister, who was also a husband and father, had a behavior pattern which he evidently could not control. He would drive his vehicle up to a schoolyard and call little girls over to his vehicle and when they got in position to look inside his vehicle, the minister would expose himself. The little girls would be terror-stricken and would run away screaming. The minister would apparently get enjoyment out of committing this act. This occurred on several occasions. Finally, a school official reported it to Mike. Several people offered information about various vehicles they had seen hanging around schools and Mike developed a suspect. Mike located a picture with the minister posing for a photo with several local people. He took the picture to several schools and had the school children identify the suspect.

Some complaints were coming in from adjoining counties. Mike met with the detachment commander of Wirt County and discussed the situation with him. It was decided that Mike would arrest him in Jackson County and let Wirt County incarcerate him and then charge him with offenses there. Mike really felt bad about the whole thing, but he proceeded to the minister's home and placed him under arrest. The minister stated that he would not think of doing such a horrible thing. The offense was just a misdemeanor punishable by a fine, but his behavior had to be stopped. The minister was transferred to Maryland and the case was never tried in Jackson County Court. About six months after Mike got a call from authorities in an adjoining state wanting to know about the minister's background. He had neglected to change his address on the registration card. He was under surveillance by police for the same offense. Mike heard later that he had been transferred to New York state and still had a church. Mike discussed the facts with a psychologist and learned that a person is never successful controlling the urge without professional help.

One night the trooper got the idea to put a dead fish under the front seat of the Fire Chief's car. In two or three days it really smelled to high heaven. The Fire Chief turned a fire wagon red and he conducted his own investigation of who did the job. All the evidence led to the trooper. The Fire Chief decided to get even. He got some ripe Limburger cheese and spread it up around the heater of the trooper's police car. The weather was cool and everyone used their heaters. It took months to get rid of the smell in the police vehicle.

In late 1961, an article appeared in Parade magazine,

exposing what they termed graft and corruption in our new interstate road system. The article began by claiming that America's new 41,000 mile interstate system was a national disgrace. The article quoted unnamed investigators that half of the highway officials should be put in jail and that graft payments amounted to an average of $100,000 per mile of the Interstate system. The article also stated that the proper people of Ripley, West Virginia rejoiced when the new highway cut through a notorious bawdy house outside of town, but the state paid the madam such a generous bribe that she was able to open a far fancier establishment. It was rumored later that certain highway officials were awarded special favors. Cpl. Mike Casey of the Ripley Detachment stated that Ripley is an extremely conservative community. Nothing like that would be permitted. The local paper took a picture of a cat sitting in the doorway of a well-built little house and said this is the only cat house in Ripley and the madam of that house is shown in the doorway. A subpoena for Jack Anderson, the writer for Parade magazine, was issued by the Prosecuting Attorney, summoning him to appear before the grand jury. He never appeared.

Mike learned to play dominoes, a favorite game in Ripley. In the wintertime things were pretty slow, especially at night. The county courthouse set aside a room for the old-timers to play dominoes and some of them were experts. Mike spent some time in the courthouse when he put prisoners in jail and he always played a game of dominoes. The old men were delighted and Mike had a lot of fun. It all came under the title of public relations. Sometimes checker-playing old men were put on juries.

Mike had a better social life than at any time in his life. He played bridge with several people, including the principal of Scott High School, and the owner of the drug store, the superintendent of Cedar Lakes, and several other prominent citizens. He attended lodge meetings and was president of the Methodist men's group at church. He was available as a speaker to any group that requested him. Mike's Air Force Reserve unit was thriving. He had two ex-service men from Ripley in his unit and provided transportation for them.

One morning Mike was in front of the courthouse talking to several people, including Gay Winters. An overbearing, retired farmer that no one liked, was complaining about the smell and attributing it to Gay. Winters was a little ripe and his colors didn't match. The farmer, who had cheated half the county on cattle deals, kept right on pushing. He told Gay he should go down to the creek and take a bath. Everyone including Gay was getting tired of the whole thing. Gay told him, "Wal, I reckon that I can clean myself up on the outside if you will clean yourself up on the inside". Mike was proud of Gay. The insulting farmer shut his mouth and walked away.

The man who owned the house that Mike was living in, decided that he should bring his mother into town. Her health was bad and she needed more attention. It was a nice place and Mike hated to move. They located a less attractive house nearer town. It soon became obvious that the house served as pasture ground for a herd of water bugs, almost as big as camels. Martha was not impressed. There were no good places to move to, anyway they were tired of moving. So, they decided to buy a house. After several months they located a place on Walter Street and moved in right away.

CHAPTER 11

Huntington

Mike completed the requirements for the Armed Forces Industrial College by correspondence while in Philippi. He completed the Command and Staff College by correspondence and two weeks in residence at Maxwell AFB, Alabama. This activity made him eligible for promotion to lieutenant colonel. All he needed was time in grade. The new law regarding military leave was working well, but Mike was still the only reservist in the department. During the summer of 1964, Mike was transferred to Huntington as the new commander of the Huntington District. He received appointment as Chief of Cabell County Civil Defense Police unit as a Reserve assignment.

When Mike was transferred to Huntington, he came in contact with a long time boyhood friend. The first time Mike remembered Sam Watkins was when he was about five years old. He was with a group of neighbor boys swimming in the Guyan River. Both Sam and Mike swam like otters. They swam the treacherous Guyan River when it was at flood stage.

Sam Watkins had also chosen a career as a policeman and was Chief, Detective Division in the Huntington Police Department. Sam was interested in making Huntington a better place to live. He got a good response from Mike in this regard. They worked together well and solved some law enforcement problems. They were both very prominent in Air Force Reserve affairs and attended training sessions together. Mike considered Sam his closest friend and Sam was in charge of the Air Police unit in Huntington and Mike was commander of the Charleston unit. Both were past Presidents of the Reserve Officers Association.

The Huntington barracks was a comparatively large facility, also several pieces of equipment were assigned. Mike tried to keep his troops busy with criminal activities. He usually had a trusty around to wash cars and take care of other equipment. He made periodic trips to the Medium Security Prison to get a reliable trusty. This usually resulted in an overnight trip. Mike would interview three or four prisoners before he found what he wanted. Mike usually followed the same routine. He would arrive at the prison late in the afternoon. He had already arranged to interview four prisoners. He never considered a check artist or sex criminal or a thief. They always spelled trouble. He would ask for and get copies of the prisoners' files. He scanned the files and talked to each prisoner briefly. This was mainly to associate a name with a face. All four prisoners were in for murder of one type or another. Many times otherwise normal people would kill someone in a fit of passion, usually they made excellent employees. A sex criminal would have too many chances to commit some sort of sex crime and would be a constant source of trouble.

The next morning Mike would conduct more lengthy interviews and make a selection. On this particular trip, he selected a trusty by the name of James Oder Jones. Yes, Oder was his middle name. However, he was a very clean person and was obviously wanting to get away from the prison. His crime was murder.

He had served about seven years of his sentence and had several more years to go and seemed like a good prospect. Mike took James Oder back to the district office and put him to work. It panned out well.

Oder was slow and would creep around, never moved fast, guess he figured he had plenty of time to get everything done. Oder had a good place to live. Sometimes they took him to the state hospital for meals, usually on Sundays. Otherwise, he did his own cooking. The troopers paid him a small sum for washing their cars. So he had some change to buy groceries. About eleven month later, Mike retired and Oder still had his job. Oder added to his income by making picture frames and tie racks. It was a good thing for all concerned.

Mike and Martha purchased a lot on Taylor Road and built a nice residence on it. Everyone, including Mike, hated to leave Ripley and the family returned as often as they could for brief visits. Mike's knee was bothering him quite a bit and he decided that he would apply for a disability because of the line of duty knee injury. In the meantime, a state police class graduated from the Academy and Mike was pleased that he had received two new troopers for the Huntington District. The new troopers were both fine young men. Mike relished the idea of getting them the proper training.

Mike was still toying with the idea of a disability retirement, but the assignment of the new troopers and the fact that he did not have a job caused him to wait a while. Mike had a lot of responsibility. There were about 15 or 20 troopers in the district and they were very busy. They had full police powers to enforce every law on the books.

The new troopers were working out well. They saluted anything that moved and it was months before they relaxed. One of the new troopers was from the Beckley coal fields, the other was from the northern part of the state. One day Tom, the unit from the coal fields, gave a lady a traffic citation for speeding. The lady had a friend who was in the Legislature and had some political power. The lady went to her friend and asked him to intervene and get the ticket fixed. The legislator called Charleston and Charleston called Mike. A ranking member of the department told Mike to instruct the young trooper to tear up the ticket. Mike had reservations about the request and he called Tom in to discuss the situation. Tom stated that the lady was traveling 75 mph in a 55 mph zone, but that he had only charged her with 69 mph in order to keep her from being assessed points. He also stated that she had an expired inspection sticker. Mike tried to make a deal. He told the officer at Department Headquarters that he would be willing to tell the trooper to recall the speeding ticket and charge her with the sticker violation. Mike also told the headquarters unit that it would have a bad effect if every young officer was questioned about his legitimate arrests. The headquarters officer said, "I didn't ask you for a lesson in psychology, Casey. I want you fix that ticket." Finally, Mike felt tired and bored with the whole thing. He sent a letter to the retirement board requesting that his eligibility for a disability retirement be discussed. In a few weeks the board was convened and Mike was awarded a retirement due to a line of duty knee injury.

CHAPTER 12

Civil Service

Mike took a Civil service test and was accepted by the Corps of Engineers as a safety assistant. Mike was accepted as a GS-5 at a salary of about $6,000 a year. This was less than his secretary was making. Mike's boss was a GS-12 and he was a good man to train under. Mike planned to lessen the wide gap in pay. He dug in and started a Motor Vehicle Safety Program second to none. After four or five months it became obvious that Mike was underpaid. The chief told Mike to take the Federal Service Entrance Test over again and they would select him off the register as a GS-7. Mike took the test over again, passed it, and was promoted to GS-7. Instead of being a Safety Assistant, he was now a Safety Officer. After one year in grade, Mike was promoted to GS-9. The Chief of Safety, was promoted to GS-13 and transferred to another division. Mike was made acting Chief of the Safety Office. Mike was determined to increase the efficiency of the health unit. He started a search for a nurse that could make it operate in an efficient manner. The person selected was almost overlooked, she was in the hospital recovering from an operation. Mike was not impressed with the other people he had interviewed. He decided to at least talk to the indisposed nurse. The interview went well. The nurse had a world of experience. She was intelligent and capable and seemed to want to develop a program on her own initiative. Mike hired the nurse and never regretted it. She made his health unit look extra good. Mike asked for and was granted permission to hire a student aide. Mike had received two outstanding performance awards and in 1969 was promoted to GS-11. The Huntington District won the Division Safety Award for the first time, but it wouldn't be the last. Mike also received a letter of commendation from the District Engineer.

About this time a very tragic event occurred. The Silver Bridge across the Ohio River at Point Pleasant collapsed and some forty-six people were killed. River traffic was down for several days. Mike spent part of the next three weeks supervising safety activities. The District Engineer was very much interested in finishing up the salvage activities without a lost time accident.

Several contracts were let with commercial firms. The area was a beehive of activity for several weeks.

About this time Mike was promoted to lieutenant colonel in the United States Air Force Reserve. This was one of the proudest times in Mike's life. The little lad who lost all his belongings in a house fire and lived in a chicken house for several months was now an Air Force lieutenant colonel.

Mike had already made contact with his old pilot and planned to visit him on the next trip to Florida. R. L. Cooper was a very capable pilot and never seemed to get excited. He usually flew with the Operations Officer. The Ops Officer, a first lieutenant by the name of Simpson had a little influence with the flight schedulers. He and Cooper usually managed to get Mike as their static-chaser. The radio operators were supposed to be selected by chance, but it didn't work out that way. There were many advantages to flying with the same crew. Mike visited with Cooper twice at his home and found out that he had stayed in the Reserves and retired with the same rank as Mike, lieutenant colonel. They compared log sheets and had a wonderful afternoon talking over old times.

On February 26, 1972, a catastrophic event occurred in West Virginia, one hundred eighteen people died as the result of the impoundment failure on Buffalo Creek in Logan County. The Corps of Engineers was given the task of cleaning up the area. Mike spent several weeks fine-tuning the safety program and was reasonably successful. Several odd things happened during the recovery phase. One day in early March, National Guard troops were still searching for bodies, they would see pieces of cloth and reach back to examine them and occasionally suffer snakebites. What happened was very interesting. The ruptured dam washed hundreds of snakes out of the cliffs and from under houses and lumber piles. In the cold weather, they had been practically immobile. On this particular day the sun came out enough to warm up the cold-blooded reptiles and about seven snakebites were reported. There may have been others not reported. The problem was discussed at a safety meeting. Corrective action was taken and it ceased to be a problem. The workers were told not to reach back into debris piles without the protection of heavy leather gloves and heavy arm protection.

The railroad was torn up and the rails were bent out of shape. The miners had been disposing of old automobiles in the creek along the bends in the river. This was to keep the river in its channel and keep down erosion. It left hundreds of old car bodies exposed. Some of them were preserved enough to use for body parts. Mike's safety efforts did not go unnoticed and he received a letter of appreciation from General Stames.

Mike's oldest daughter, Saundra, graduated from Marshall University about this time and decided to follow her father's footsteps and seek an Air Force career. She applied for and was accepted into Air Force Officer Training School. Mike's youngest daughter, Martha, was married and stationed with her Air Force husband at Rickenbacker AFB in Columbus, Ohio. Sandy entered the OTS program in the pay grade of a staff sergeant and after a rather hectic time graduated as a second lieutenant. Mike had a meeting of the National Water Safety Congress about this time. He was the executive vice president of the organization and the program was dependent on his attendance. The meeting of the National Water safety Congress was in Dallas. Mike and Martha did a balancing act, they drove to Dallas where Martha dropped Mike off for his meeting. Then Martha went on to San Antonio where she pinned brand new lieutenant bars on her daughter. Mike was very disappointed that he could not attend this very significant occurrence in his daughter's life. He announced it to the congress. Later that afternoon Martha brought her to Dallas, bars and all. She was stationed at McCoy AFB in Orlando, Florida. During the winter, Mike and Martha took Martha Louise to Florida with them. They visited Sandy at Orlando and spent a lot of time at the beach. In November 1973, Sandy married a B-52 pilot and moved to California. Sandy's husband, Rod, was stationed at March AFB and Sandy was stationed at Norton AFB.

In early 1973, Mike was appointed as the military counterpart of the State Director of Civil Defense. He never missed a training session and worked hard at his job. Several disasters occurred during this time frame in which Mike was heavily involved. Mike had three tours of duty in Battle Creek, Michigan and completed all the staff college courses.

Mike had been a member of the Reserve Officers' Association for about 30 years. He had been a president of the West Virginia Department and had served as chapter president in Huntington for several years, Mike was a member of the Odd Fellows Lodge and had served as a Deputy District Grand Master for two years.

Mike had never forgotten Kenney, his old buddy in radio school. When he went to work for the Corps of Engineers, he put himself in a position to travel to Cincinnati. During the summer of 1966, he spent two days there. The next time he was there in the summer of 1967, he contacted a man who had a first cousin named Charles M. Kenney, who had served in the Air Force during World War II.

Mike called Charles' number that night and it was indeed his old buddy. They arranged for Kenney to pick Mike up in front of his hotel the next afternoon. At the appointed time, Kenney drove up in a big car, dressed in a very handsome suit of clothes, and took Mike to his home. He had five or six children and was president of a small manufacturing company. Neither of them knew where Shag was, but had heard from someone that he was still in the Air Force. The next time Mike went to Cincinnati, Kenney's son advised Mike that Kenney had died of a massive heart attack two months before the visit.

Mike had been acting Chief of the Safety Office for quite some time. They still could not make him the chief because of the engineering requirement. The office of Chief of Engineers requested that a safety professional be selected to head up the safety effort at Langdon, North Dakota, the site chosen for the anti-ballistic missile complex. Mike was chosen. This was just the kind of exposure Mike needed. He was given the opportunity of administering a safety program with a variety of activities involved.

Orders were cut to send Mike to North Dakota. He decided to take his family with him. They didn't lose any time. They closed the house for at least a month and headed out for Chicago. They stayed all night in Chicago and the next day they went sightseeing. The following day, they reached Grand Forks and then on to Langdon.

When Mike arrived they were constructing silos for the sprint missile and had already suffered several lost time accidents. The project kept about 2,000 workers busy at several locations. The largest project was the radar itself. It was a massive concrete structure. The project was called the Safeguard Ballistic Missile Defense System. The safeguard was designed to fulfill three objectives: first, protection of our land-based forces against a direct attack by the Soviet Union; second, defense of the American people against the kind of nuclear attack which

Communist China was likely to be able to develop within the decade; third, protection against the possibility of accidental attack from any source. The president emphasized the defensive nature of the system. The safeguard weapons system consisted of five basic components: Two types of missiles, two types of radar, a computer and data processing center. The radars were interconnected by the computer network. Spartan was the long-range punch of the safeguard. It was fairly large, 55 feet long and had a range of several hundred miles. It was designed to fly very high to destroy inbound enemy missiles. The system was probably obsolete before Mike arrive at the site, but everyone was working as if their lives depended on it.

The sleepy little town of Langdon was not geared to handle the sudden influx of people. More than five thousand people were involved, including dependents. Langdon's normal population was probably 3,000. The public utilities system was quickly overwhelmed by the increased demand. A trailer park was constructed to house several thousand people. The traffic situation was intolerable. Since several components of the missile system were located near Langdon, the town automatically became a target. This caused resentment and much grumbling from the local population.

The Perimeter Acquisition Radar System (PARS) was the long-range target acquisition radar. As the name implied it was located around the perimeter of the defended area. In case of a safeguard deployment, PARS were located across the northern boundaries of the nation and as appropriate on the East and West Coasts. PARS had a range of more than 1,000 miles and could detect an intercontinental ballistic missile as it rose above the horizon. It was to track the incoming missiles to determine their speed, direction, and probable targets. This information was then passed on to the computer at the Missile Site Radar (MSR). The MSR was at or near the missile fields. Its mission was to track the incoming missiles when they got within range and also to track and control defensive missiles. In a nutshell, PARS would provide the warning and MSR would fight the battle.

There wasn't much to do in Langdon. They had one movie theater and one golf course. The golf course had sand greens and after a player putted out he had to rake and smooth out the greens. Once a week they had a steak cookout at the golf club. Almost everyone attended. They bought up several cuts of steak, enough for two or three hundred people and had them on display. The hungry patron selected his steak and went to one of the many charcoal grills and cooked his steak. It worked well. After the steak, they had several poker games going at one time. The steak fry was well-attended and everyone enjoyed the weekly event.

Mike gained a lot of experience in fire prevention. He helped organize fire departments in the areas nearest the projects. He assisted with fire drills and attended meetings with city officials.

The town of Langdon was not too far from the Canadian border. The lakes in the area were teeming with Great Northern pike. Mike went fishing several times and caught his share of the fish. The family ventured into Canada and on one occasion they visited Winnipeg. It was a beautiful city and the nomads had a wonderful time.

The family returned to Huntington. Mike stayed an extra two or three weeks fine-tuning the safety program. Then he too left for home. He flew in from Grand Forks by commercial air. The project engineer was very lavish in his assessment of Mike's performance. He wrote a very complimentary letter about the safety activities to the Huntington District Engineer.

Mike had several poker sessions at his home. He would invite the district engineer, chief of construction, chief of operations, and two or three others. Martha always cooked beans and cornbread with a big slice of onion for a snack. The chief of construction said, "Casey was the only man in the world who had the nerve to invite the district engineer to his home and feed him beans." The chief of construction said his beans always cost him $10.00 a bowl and he ate two or three bowls. An invitation to a Casey bean dinner was considered a social plum.

In 1979 another attempt was made to put Mike in charge of the office. The regulations still called for an engineer to occupy the position. Mike returned to Huntington as a Safety Specialist. He saw the handwriting on the wall and decided to get the position for his assistant who was a graduate engineer. Mike was successful and his assistant was made chief of the office. In late 1979, the Chief of Safety asked for and received an appointment to another agency. The District Engineer again requested a waiver of the engineer requirement in the Chief of Safety position. This time the requirement was waived and Mike was installed in his long sought for position. Mike needed to select an assistant and he decided to support the EEO program and appoint a woman to the position. He interviewed several women for the job and selected a young lady with a Master's Degree in Safety Administration. What she lacked in experience she made up with enthusiasm. She was really an asset and again the image of the Safety Office was enhanced.

Mike began to wonder about what had happened to some of his old buddies in the service. He contacted one of Shag's brothers and learned that he was retired from the Air Force and lived in Oklahoma City. Mike wrote him a letter. Shag answered and immediately gave Mike a phone number. Mike called and learned that Shag had retired from the Air Force. He was in poor health. Shag made a trip to Huntington to visit his mother. While there he spent an entire afternoon with Mike. Two years later he was diagnosed as having Alzheimer's disease and was dead within a year of the diagnosis.

Mike had belonged to the NWSC for over sixteen years. For the last fourteen years he had been a regional vice president. In 1979, he was elected executive vice president. Mike handled the funds in a Coast Guard grant to produce a water safety film. Martha helped with the bookkeeping and the film was produced. The production was a success and Mike received a President's Citation for it.

CHAPTER 13

Civil Service Israel

From 1978 to 1980, three big-timers were prancing across the world stage. Several bit players were lurking in the wings. But! Menachim Begin, Jimmy Carter, and Anwar Sadat were the stars. They bounced around between Egypt, Israel, Camp David. They studied their scripts well and were always just on the verge of establishing a permanent peace between Israel and the Egyptians. The three stars all had tremendous egos and played their parts well.

The most critical part of the negotiations was the return of the Sinai to the Egyptians. Israel won the Sinai as a result of a very costly war, suffering casualties that they could ill afford. The largest consideration was that Israel needed the desert as a buffer zone between themselves and a traditional enemy. Israel was left with very little air space to maneuver in. Once a fighter plane took off it was out of the country in a matter of minutes. They had given up a lot of air space when they returned the Sinai to Egypt. Pretty soon the peace accords became an established fact. An agreement was finally reached. The Egyptians were given back the Sinai. The Egyptians made some concessions. Free passage through the Suez Canal was one consideration the Egyptians granted. The agreements included two air bases that the Israelis ceded to the Egyptians. The U.S. promised to build two state-of-the-art air bases for the Israelis to replace the ones given to Egypt. The U.S. started building the new bases. The State Department and Corps of Engineers began to get construction fever and started planning for a new border between Egypt and Israel.

The new bases cut the turn around time for getting a fighter plane back into the air by at least one half. The facility was designed to get a fighter into a secure position underground, tend to the pilots' physical needs, refuel the aircraft, and rearm. Minor maintenance adjustments could be accomplished to get them back in the air quickly. Reducing the turn-around time has the same effect as adding new aircraft to the inventory. The Israelis were one of the world's best in getting the most out of an airplane. They were probably getting three times the number of sorties out of one plane than the Palestinians or Egyptians.

In December 1981, the Office of the Chief of Engineers requested that Mike be assigned to the Middle East as chief Safety and Security Officer. The State Department was involved with bases and observation posts for the peace-keeping forces. Posts were to be built along the new border at strategic intervals. The project involved the construction of two main bases. One in the north and one in the south. About 30 small command posts were to be constructed in between the two bases. These command posts in general followed the new border between Israel and Egypt.

The command posts were equipped with a generator and necessary communications equipment. The trailer-like living units could take care of two or three men very well. Mike was instructed to be in Washington the next afternoon to be briefed and to get a passport. Arrangements had already been made to obtain a passport in one day and get back home that evening. Mike accepted his assignment and received his passport. He was to leave for Israel two days later.

The travel orders provided for Mike to fly commercial air to Orly Airport in France, then on to Tel Aviv, Israel. Mike was to stay two weeks and return home before Christmas. Mike was to prepare plans for a construction safety program and get the materials and supplies he needed to pursue a vigorous safety campaign. He was to return to Israel in late January 1982. Mike was an old hand at getting the house in shape to sit idle while he and Martha traveled. But this would be his longest trip and he didn't have much time to get ready. Martha had to pay bills ahead of time and get things ready before she left to be with Mike. This was the start of one of the busiest and most exciting times in Mike's life.

Flying was a little dangerous. The airlines were having security problems. Several highjackings had occurred. On the flight from France to Tel Aviv, a disturbance occurred and Mike provided some assistance to a stewardess. An intoxicated Palestinian started behaving in a belligerent manner. Mike talked to the stewardess and informed her of his State Police background. The drunk was getting loud and aggressive. At that point the stewardess requested Mike's assistance. Mike got a head lock on the passenger and put him down in his seat and held him until he settled down. The man finally dozed off in a drunken stupor. The pilot later thanked Mike for his assistance.

When the plane landed, the first thing Mike saw was two soldiers with uzis. A man employed by the Sinai Construction Management Office (SCMO) met Mike and assisted him through customs. The man had a vehicle which had been assigned to Mike. The vehicle had State Department license tags on it and had almost unlimited parking privileges. Mike could park it at the airport or most anywhere else at no cost and not be concerned about it. SCMO Headquarters was in the Palace Hotel where Mike lived. A security gate kept traffic out of the hotel compound, which provided parking for official vehicles. Mike observed all this with interest as it would be his responsibility to supervise security activities.

The people who were responsible for constructing the two air bases were also housed in the Palace Hotel and had been there for about two months. The hotel was formerly a luxury hotel and was surrounded by other resort hotels along the Mediterranean Sea. Mike was assigned a room and noticed that the size of the room and the degree of luxury was determined by the rank and/or grade of the individual. His Air Force rank wasn't considered in room assignments, so he had to concentrate on Civil Service grade. Mike decided right away to maneuver for a Grade 13, which put him in a different category as far as perks were concerned. In Civil Service the number of square feet in one's office is determined by grade. It may be the difference between having carpeting or not. It made a considerable difference when it came to pay.

There was a huge dining room where Mike had all his meals when he wasn't in the field. The dining room could accommodate probably 200 people. The food was outstanding and the service was good. The waiters were provided by a nearby school for waiters. They were all young and well-trained. Israel produced some excellent fruit and citrus. They had an adequate dairy industry and produced some good dairy foods. Eight flavors of ice cream were in one corner of the dining room on a self-serve basis. Mike would have to watch his weight.

The office was adequate for what Mike planned to do with it and it was near the head shed. The Safety Office had a generous travel allowance and Mike went to the field the very next day. Visits were planned for the entire project so that Mike could get a feel for his activities. Ten countries had volunteered troops to help with the peacekeeping—Australia, Great Britain, Columbia, Fiji Islands, Italy, Netherlands, Norway, New Zealand, USA, and Uruguay. The troops from the Fiji Islands were the sharpest of the lot. They were well-equipped and well-trained.

The construction projects included two main installations. One at Sharm El Sheikh and one at Eitam not far from Gaza. Headquarters for the forces were at Eitam. They were referred to as the north and south sites. Eitam being north.

Mike covered as many of the construction sites as he could and visited Jerusalem and Bethlehem on the weekend. He wanted to see as much of the country as possible in case he didn't get back. Mike put in a lot of overtime on the projects and had a healthy paycheck coming in every two weeks. He spent two days driving the entire length of the Sinai and visited several of the observation posts. Mike managed to get to Masada and Jericho before he returned home. He was allowed generous mileage on his vehicle and had no trouble obtaining gas. He ran into several situations where unexploded ordinance was a problem. Finally, the busy two weeks quickly expired and it was time to go home for Christmas.

Martha had filled the entire garage door with a poster, which read "WELCOME HOME FROM ISRAEL". It made Mike feel that he was wanted and appreciated. Mike took one day off and showed up at the office the next day. He immediately went to the personnel office and requested them to reclassify his job and make it a temporary GS-13. He would worry later about making it permanent.

He had two weeks to prepare accident prevention plans. Mike started gathering personal protective equipment and other safety items. He selected several films for training purposes.

Personnel agreed that the position was a high visibility assignment and that they would take a good hard look at it. The time passed quickly and soon it was time to return to Israel. Just before Mike left the District Engineer personally presented to Mike his promotion papers, making him a G-13.

There was much to do. Mike planned to get back to Israel late in January. Martha planned to leave in early February. Mike had a friend in the operations division who had been involved in the Safety Program and was willing to work with Mike again. He was a graduate engineer with plenty of smarts. Mike arrived back in Israel in the middle of January, eager to work in his new grade. The weather was beautiful. Mike was assigned very nice quarters. The promotion made a subtle difference in Mike's working and living conditions. He was told that they would reassign him when Martha arrived.

Mike started his tour by traveling to Sharm El Sheikh to conduct safety surveys and participate in a pre-construction conference. The construction companies had their own safety program with offices both north and south. Each office was managed by a retired U.S. Army Master Sergeant. The surveys went well and Mike was encouraged by the enthusiasm of the safety personnel.

Martha was due to fly over pretty soon and Mike anxiously awaited her arrival. Mike had contracted with an Israeli citizen, who knew his way around customs, to get Martha through without the usual long delay. Martha was processed and checked out of the building in nothing flat.

Mike drove Martha to the Palace Hotel located on the Mediterranean Sea and right in the middle of several other resort hotels in Tel Aviv. Arrangements had already been made to get Martha some meal tickets. This enabled her to have three meals a day for as long as she was there. The room was really nice and Martha had time for a nap before dinner. After an outstanding dinner, people came by to welcome the newcomer. Mike spent the rest of the week working in his office and showing his wife the town.

A Jewish man, who was a good friend of Mike's, had a son in Israel. He was an artist and lived in a village called Ein Hod. One afternoon Mike and Martha drove out to Mount Carmel to visit the artist colony. His friend's son made his living producing stained glass. Mike spent a pleasant afternoon there.

One weekend Mike decided to visit Masada again. Martha had been wanting to make the trip and another lady who worked for SCMO was interested. Mike had made the trip during his first two weeks in Israel and had taken the ramp access that the Romans had used for the final assault and walked to the top. The trip took them by Jerusalem through the Judean Desert and down to the Dead Sea Valley which is 398 meters (1233 ft.) below sea level and is the lowest point on Earth. Their first stop was in Jericho, the oldest city in the world which was 820 ft. below sea level. From Jericho they drove south along the shores of the Dead Sea to Ein Gedi, an oasis in the Judean Desert. The springs and water falls were beautiful when compared with the shoreline of the Dead Sea.

They arrived in Masada in the afternoon and rode the cable car to the top of the mountain. Once or twice a year Masada becomes the scene of a exciting and symbolic event. Regiments of young soldiers of the Israeli Defense Forces fresh from basic training take the solemn oath of allegiance to the state.

The three also visited the town of Beersheba, which had never been completely excavated yet. Modern Beersheba continues to amaze the tourists. The market is an event that takes place once a week. Here Bedouins from all over the Negev meet to trade and barter camels, sheep, clothing, produce, and household goods. They have been doing this since time immemorial. The Bedouins hail from Saudi Arabia. They are a nomadic people and their black tents, made from woven goat hair protect them from the burning sun, violent winds, and sand storms. It turned out to be a very exciting day.

The Israelis were experts in the use of water resources. The Sea of Galilee was the only source of fresh water in the country. All of the water from the Sea of Galilee ran into the Jordan River and on to the Dead Sea. There was very little left when it reached the Dead Sea.

Mike had a lot of security problems in the Sinai. One of the outposts was located on top of a mountain. They needed to get a generator and other equipment up to the site and the road crews had not yet completed a road to the area. They decided to get a helicopter capable of handling external loads. They flew the generator up to the site and left it overnight. No guard was put on the equipment and the next morning it was a mess. All of the hoses and wires were pulled out of the equipment. The Bedouins visited the spot that night. They put sugar in the oil and gasoline tanks, wrecked the carburetor and just literally ruined a brand new generator. This set the construction schedule back for several days. The generator was to have been used as a power source to complete the project. The Bedouins didn't like the idea of strangers running around their desert.

Security was always a factor at most places where people gathered. They had security pits. Metal-lined holes in the ground, with a metal lid. Packages, handbags, or other suspicious articles could be placed inside the pit and detonated. These security pits were in all public areas such as airports, sports arenas, etc.

The observation posts presented a lot of problems. The living units were pre-fab and there were antennas and communications equipment everywhere. At some locations they could not get a ground on the generator. The Sinai was very rocky and it was impossible to drive a rod into the ground. When you dug a hole for the rod you could not get an adequate ground. At some locations they let all the waste water run over the grounded rod into the hole.

The outposts were very lonely places. Troops were rotated every few days. They needed to establish occupancy immediately after construction or put guards on them. Reliable guards were hard to find. Living quarters for the commander, usually a general officer, were constructed at Eitam Headquarters for the peacekeeping forces.

All sorts of minor incidents were occurring that could prove to be embarrassing. An accident that took place near Eitam is typical. The narrative report of the incident was as follows: On 12 April 1982, at approximately 1700 hours, Mr. Prankit Phoongern, employee #353, a Thai carpenter driving pickup #MFOO75 from checkpoint three, struck and killed a goat belonging to a Bedouin tribesman (name unknown). The goat was one of a herd that had crossed the road and suddenly jumped back in front of the truck. Prankit stopped the truck and tried to contact the goat owner, but the owner ran away. Prankit then drove back to the Eitam site and reported the accident to the Thai security chief. The value of the goat had not been determined. Recommended preventative action: the driver should be cautioned on the necessity for defensive driving and the goat owner should be reimbursed. This accident could hardly be considered an international incident, but it could cause resentment and lead to sabotage or destruction of property.

When Mike first visited Eitam, he passed a tell about ten or twelve miles from Gaza. He stopped by the tell every chance he got. During the First World War, British forces, composed mainly of Australian divisions, camped around Tell Jamme while preparing to attack Gaza. Of the thousands of shells which exploded, there was one which fell along the Besor Brook and uncovered a mosaic tile floor which was transferred to Australia after the war. The tell was several thousand years old. It was a very interesting place. A road had been cut through a slight rise and pieces of pottery could be seen embedded in the earth along the cut. Mike spent some time there whenever he could.

Mike did a lot of digging around the tell, and uncovered some large pieces of broken pottery. Many pottery shards were uncovered. The handles of an earthen jar were the most durable parts of it and are often dug up unbroken. Mike got enough of the handles to provide napkin rings for a six place setting.

A tell, described in simple terms, is a mound of earth built up over hundreds even thousands of years. It is from necessity on comparatively level ground where the sand or soil can drift, blown by the winds, to an object, usually man-made, which collects the sand. As the sand and soil collects, the mound gets larger. After many centuries have passed, it becomes a very interesting place. Quite often several layers of civilization may be covered up at one tell. Sometimes a tell takes in a good sized village or town. Antiquities are so common in Israel that sometimes they are used for building materials. Much excavation remains to be accomplished.

Tell Jamme was apparently used to store large amounts of grain as well as other staples. The tell covered two or three acres of ground and was about 80 to 100 feet high. One pretty day Mike decided to go to Eitam on business and he decided to go through Gaza, even though the town was known to be dangerous and had been off limits for some time. Mike and Martha went their merry way through the town. The town was crowded with early morning shoppers and commerce. The people were very unfriendly. The drive took about fifteen minutes and Mike breathed a sigh of relief when they left the area.

Mike and Martha drove to the tell, which was on the road to Eitam, to look around. The tell was fifteen or twenty miles from Gaza and Mike stopped to show it to Martha. Some Israeli soldiers were on a training exercise and had some communication equipment on top of the tell. The height improved radio transmissions considerably. The soldiers were glad to see them and told them the history of the tell. Mike explored the area to the rear and learned that a huge piece of it was caving in. Certainly, many treasures would be found as erosion continued. Road construction left pieces of pottery exposed. The tell had four or five cistern-type vats. It was apparent that the tell was a storage place for grain among other things. Mike and Martha went on to Eitam where the main peacekeeping installation was located. Some buildings were being modernized and repaired. Others were being constructed from the ground up.

Mike and Martha decided to make a trip to Egypt and obtained space on a guided tour. About 25 people were on the bus, five Americans, ten Israelis, and ten Palestinians. The group went through the town of Yemet, crossed the Northern Sinai, and pressed on to Egypt. The tour bus stopped at the border for a customs' check. It took about two hours. They went through the Israelis' luggage with a fine tooth comb then scattered it around and didn't put it back. They never looked at Mike and Martha's luggage because of the State Department passports. They crossed the Suez Canal on a barge. It was quite an experience watching the huge ships materializing out of the desert. From where the bus was Mike and Martha could see the ships, but couldn't see the water. They settled in at Cairo and spent the night in a nice, quiet luxury hotel.

Seeing Egypt was a wonderful experience. After a good meal and a night's sleep, Mike and Martha were ready to go. First they visited the Sphinx. Then they mounted camels to ride up the slight hill to the pyramids. Martha's camel could not control itself and defecated heavily. Mike's camel slipped and fell to its knees. Mike could not keep his seat and he slid over the camel's neck and landed in a pile of fresh camel dung. It was soft and squishy and smelled to high heaven. There were no rest rooms at the pyramids, and Mike had to suffer in a smelly silence. He had to wear a soiled foul-smelling shirt, soaked with camel poop, the rest of the afternoon.

The group went on to the pyramids. Mike and Martha went back in one of the rooms to view some inscriptions on the walls. It was a small room with electric lights strung around. The lights had gone out and the attendant had lit some kerosene lanterns. It was a very dangerous situation. The room was small and the lanterns were burning up all the oxygen and the air was fast becoming dangerous to breathe. People were becoming alarmed. Mike grabbed Martha by the arm and got her out of there. The bus went down to the Nile. It stopped long enough for Mike to pay a girl a dollar to wash out his shirt. She was very pleased.

While traveling along the Nile, the bus met a truck coming from the opposite direction. It was loaded with cattle. As they passed, one of the cows stuck its head out and hit the side-view mirror on the bus. It almost decapitated the cow. Teeth, bones, and blood flew back through the open windows and hit several people, including Mike and that made some more smelly, messy materials to contend with. It was time to quit and they returned to the hotel. It was really nice to clean up and get a good night's sleep.

The next day the group went to the Cairo Museum and spent a very interesting morning. They recrossed the Sinai and headed for Tel Aviv. They passed sections of desert that were literally filled with tanks, trucks, and other vehicles of war. No doubt many of them still had ammunition and other unexploded ordnances inside. A junk dealer would become wealthy if he could get access to all the scrap iron in the desert.

The group arrived back in Tel Aviv in mid-afternoon. Mike had a memo on his desk advising him to be in Sharm el Sheikh the following day for a pre-construction conference. He went by commercial air that same evening. The Israelis had caused a new town to be built on the border of the two countries. Construction was almost entirely of concrete. It was built by Israeli settlers and was quickly filled. The Egyptians objected to the town so close to the border. They threatened to ignore the peace accords if Israel did not give up the town. The settlers refused to move. Israeli troops finally went into the settlement and dragged the settlers out. They then bulldozed the homes down around the settlers. Israeli troops completely destroyed the town, leaving no possibility of restoration. Israeli troops actually cried as they forced their countrymen to abandon their homes. Mike saw the town just a few hours after the raid, and the destruction was total.

The Palace Hotel was a lively place. It had a bar where drinks were sold at a reasonable price. It had a movie theater where good movies were shown. A bridge game could usually be found in the lobby. The most interesting aspect of the activity was the people involved. There were people from all over the world. The contractors hired many female Brits to do secretarial work. Many of the workers were from Thailand. Mike met a Eurasian, who was born in Chittagong, the Burmese town where Mike spent a week flying food to British troops. There was a priest in Tel Aviv who spent a lot of time at the hotel. His name was Father Phil. Mike and Martha went on several trips with him. On one of the trips to Jerusalem, they had the very rare experience of seeing snow in the town.

It was nearing the time when Martha would return home. Mike wanted her to see Masada. He planned a trip by motor vehicle. They went to En-Gedi probably one of the most beautiful places in Israel. There was plenty of water and plant life was abundant. They went on to Masada and spent several hours there. Masada was the last stronghold which held out against the Romans during the Jewish revolt. The fortress was so impregnable that the Romans never overcame its fortifications. At the end of three years the defenders put themselves to death rather than fall into enemy hands.

Mike and Martha traveled along the Dead Sea and stopped several places. They then went to Jericho. The town of Jericho was built on an oasis and some citrus fruit was grown there. Two or three hundred small refugee homes were crowded together on the outskirts of town.

It was time for Martha to board an El Al plane for home. Mike followed in about three weeks. They had a reception for him at the hotel just before he left. Little did he realize that in four or five years he would return with his entire family.

Mike came home the last of June and again settled into retirement. The District gave Mike a retirement party that was well-attended. He received several letters of commendation, some of them were heart-warming. The District personnel took up a collection and bought him an excellent 35 mm camera, which he put to good use in Alaska and Korea. Mike worked at chores around the house and got in Martha's way. He had vague misgivings about his health. But! He was in seemingly good health, still jogging three or four miles a day and going to the YMCA six days a week. He pushed the slight concern he felt into the back of his mind and forged ahead. Mike and Martha went to Florida for two months in December 1982 and returned in February 1983.

Chapter 14

Civil Service Alaska

Mike had been to Alaska way back in 1947 and had always wanted to return. He had been with a crew assigned to Fairbanks, Alaska for cold weather tests, checking the performance of different fuel mixtures. His first visit had been almost forty years ago. When he heard of the Grizzly Program and he decided to participate. The details of the program were sent to all Corps of Engineers District Offices. The program involved getting some specialists in their fields back in the saddle. They needed some well-qualified people for the short construction season. Mike decided to throw his hat into the ring. Mike sent an application to the Alaska District Personnel Office. The personnel office received over 100 applications and hired fourteen of the individuals who had applied. Ex-chief of Safety, Mike Casey, was one of the fourteen grizzlies selected.

Just before being selected, Mike and Martha had purchased round trip tickets on AMTRAK for Albuquerque, New Mexico, to visit their youngest daughter and son-in-law. Their son-in-law was a first lieutenant in the Air Force. They decided to go ahead and make the trip since no firm dates were set for Alaska. They left immediately for Albuquerque. About three days after they arrived in New Mexico, they received a call from the Chief of Engineers' office, authorizing travel to Alaska and instructing Mike to be there within ten days. Mike and Martha scurried back home and got things ready to leave. They had plenty of practice doing this. They arrived in Anchorage by commercial air, well within the ten day limit. Mike and Martha were met at the Anchorage Airport by Marty Holba, safety officer for the district. Marty's training was mostly in fire prevention and he had an electrical engineering background. Marty was married and had two beautiful daughters. Marty's wife was a warm, friendly person who managed her household well. She had a freezer full of moose meat, filleted King Salmon, and Halibut. Mike and Martha quickly became part of the family.

Marty took Mike and Martha out to the air base and Mike was in his element. He secured his quarters through the Corps of Engineers and was able to get really good quarters through his

Air Force affiliation. They had a four room apartment directly across from the Officers' Club. Their dining room was large enough to hold a staff meeting in. The apartment was equipped with blackout blinds due to the long periods of daylight. In the summer, the sun would shine over twenty hours a day.

They had an excellent Commissary on base. Fort Richardson was adjacent to Anchorage and had an excellent Officers' Club also. The Non-Commissioned Officers' Club had the best food on the base and that is where Mike and Martha hung out when they were hungry. There was a really fine base theater nearby. Bingo and other forms of amusement were available. It was beginning to look like an excellent tour.

Since it was getting late in the day, Mike decided to get a good night's sleep before reporting. Three of the grizzlies were assigned to Anchorage, Mike being one of the three. Mike picked up his assigned vehicle at the motor pool and parked it at District Headquarters. There were several people at the District Office that Mike already knew.

Mike was given a small office to operate from. The Alaska District Corps of Engineers was headquartered on Elmendorf Air Force Base adjacent to the city of Anchorage. The district has jurisdiction throughout the state on military construction and real estate transactions. Approximately 450 civilian employees were engaged in the mission of the Corps of Engineers.

Anchorage is the economic capitol of the state and had a population of 250,000. The area was similar in temperature range to the northern Great Lakes region. The shortest days during December have approximately five and one half hours between sunrise and sunset. The longest days in June never really get dark. Quite an adjustment is necessary to regulate sleep habits. Blackout blinds were used quite frequently. As in most cities of comparable size, Anchorage offered a full compliment of civic and social organizations. Workers in Alaska never seem to suffer due to lack of activity. If nothing else, you can always go fishing.

Mike started his travel right away. He went to Cape Newenham, and King Salmon. He usually traveled in a twin-engine Otter operated by commercial airlines. The Otter was a very reliable plane and could land on a short runway. A traveler might find himself flying with a dog sled team or snow mobiles, chain saws, boat motors, two or three snowmobiles, five or six

Eskimos, and other assorted pieces of equipment. The landing strips were crude and usually built on the side of a mountain. Many places the aircraft had to land uphill.

There was plenty of entertainment at the larger construction sites. The construction sites usually had an abundance of recreation equipment, including fishing rods and reels and Mike loved to fish. At one project site the river ran directly behind the quarters and Mike took advantage of the fact. He caught a fish called Dolly Varden. He caught so many that his arms got tired.

The Safety program at remote locations left much to be desired. The construction season was short and created a sense of urgency which resulted in taking unnecessary chances. In most cases, the season was over in October. Construction crews committed unsafe acts, often used unsafe equipment, and failed to use personal protective equipment. Mike designed his program to eliminate these discrepancies.

Mike and three of his friends went fishing near Kenai. They rented a 35 ft. boat and fished for Halibut. They had good luck and caught several fish, one weighed 85 pounds. They shot it in the head with a .410 gauge gun in order to get it in the boat.

Dizzy spells were becoming more frequent in Mike's daily life. One morning he was driving to work and became terribly dizzy. He drove straight to the base hospital instead of going to work. The medics were worried too and gave him a thorough physical. They decided he had inner ear problems. They gave him some pills and let him go.

Marty Holba had a friend who had just obtained his pilot's license. The friend, Ken Peters, was an engineer and a key man in the district office. Marty and Ken were less than half Mike's age. Mike could have easily been their grandfather. They spent a lot of time together. Volleyball games during lunch hour, fishing, playing cards, and many heated discussions about the construction projects occurred.

One afternoon just after Ken got his pilot's license, the three decided to go fishing for king salmon. The fish had just started their annual run up stream. Ken called the flying club he belonged to, and found out that he could rent a three place aircraft for the rest of the day. They left Anchorage about three o'clock in the afternoon. They flew about 75 miles south to an abandoned airstrip that had been used in the oil exploration

Alaska fishing trip

business. About three miles from the airstrip was a river which was never more than 40 ft. wide. The river was small enough to wade across with waist-high waders. All three of the anglers had waders, and plenty of salmon eggs for bait. They waded across the river and found some level ground on the other side. Mike stayed where they had crossed the river while Ken and Marty walked down the river for about 50 yards. Mike put a big glob of salmon eggs on his hook and let it float down the swift cold river. Wham! Something took his bait and he reeled in about a 17 inch rainbow trout. He caught two more in quick succession and threw them on the bank. He was disappointed that they were so small after hearing about 50 or 60 pound kings. Marty told Mike he had just caught three of the tastiest fish in the world rainbow trout. Marty told Mike to come on down where they were and the water was deeper. Mike complied and saw that Marty and Ken had already caught two 25 or 30 pound kings. They fished for about one hour and all together they caught 27 kings, the largest of which weighed 53 pounds. They also caught several more trout. They kept several of the kings and filleted them and cleaned them right on the spot so they wouldn't have the extra weight to carry back to the plane. They cleaned and filleted two trout for each one of the three men.

They started back across the river with the load of kings and trout. They were going to be overloaded so they limited themselves to about 25 pounds of fish per man. They threw all the other fish back as soon as they caught them. As Mike crossed the stream, he felt dizzy and looked down at the cold water rushing past. He fell into the icy cold water. This really worried the boys. They helped Mike onto shore. Luckily, the weather was very mild and although it was about 8:30 p.m., the sun was still out. They walked the three miles to the plane in about forty minutes and took off before it was completely dark. Mike told his friends that he would see a doctor the next morning and try to find out more about his dizzy spells.

They arrive home about 11:00 p.m. and it still was not completely dark. It didn't seem possible to leave Anchorage at 3:00 p.m., fly almost 75 miles, walk three miles to the river, catch several large fish, walk back to the plane with several pounds of fish, fly back to Anchorage, and be home at 11:00 p.m. Looking back on the trip, Mike felt the plane was probably severely overloaded. The mosquitoes were so thick on the way back to the plane it was hard to see and walk. Mike was really concerned about his fall into the river. He reported to the hospital again and was told inner ear problems took some time to heal and to keep on taking the pills.

The next excursion planned by the three sportsmen was a hunt for Dahl sheep. Dahl sheep stay deep in the mountains and are difficult to find. They are a much sought after trophy. The Wrangles Mountain area was known to have some of these sheep.

Mike and Martha decided to take a train trip to Mt. McKinley National Park. They made reservations to stay in a railroad Pullman car at the park. The quarters were a little cramped and the air conditioning wasn't working well, but the bus trip into the park the next day made up for it. The trip took about six hours and they saw dozens of animals and hundreds of birds. The train ride back was outstanding. Mike and Martha had an entire passenger car to themselves.

Marty and Ken decided to delay their hunting trip for a couple of weeks. Mike and Martha realized that Martha had to think about getting back home to take care of affairs. Mike had been devoting time to his safety duties. He traveled all week and was home on weekends. Mike and Martha decided to spend the weekend traveling to Whittier and boarded a state-owned ferry at Valdez. They spent the night at Valdez and spent some time looking over the pipeline and other points of interest. Several large totem poles were observed. A ride in a very comfortable van back to Anchorage completed a very pleasant trip.

Ken and Marty planned to scout the Wrangles for sheep. The next week Martha was leaving for home and she made Mike promise not to go scouting for sheep. She was afraid he would re-injure his bad shoulder with no medical assistance available. Mike gave her his solemn promise and he kept his word. He told Ken and Marty he would take a rain check. That weekend Martha left for home.

Early Saturday morning they took off for the Wrangles Mountain area and that was the last time anyone ever saw them alive. Marty and Ken flew into the side of a mountain. They apparently flew into a box canyon and couldn't gain enough altitude to climb out. A state-owned helicopter happened to be in the area and the crew saw the crash. The crew landed the helicopter as near as they could to the downed aircraft. The two young engineers were killed instantly. The bodies were taken to their home towns. Mike did not find out about it until Sunday afternoon. He was grief stricken, almost as bad as losing a son. Mike was saddened by the accident, but was thankful that he had kept his promise to Martha.

Mike kept the program going for another two weeks. The District Engineer offered the Chief of Safety job to Mike on a full time basis. Mike was tempted, but an uneasy feeling that things were not right in his life persisted and he decided to leave for home. Mike had about 40 pounds of frozen king salmon to take home. The Division Engineer requested that Mike stop off in Seattle and brief them on what was happening regarding the Safety Program. Mike was glad to comply because he had some friends in the division office. Mike had no trouble getting the frozen salmon on board the aircraft. He put the salmon in a friend's frozen food locker while in Seattle.

Mike arrived home. He was worried about his lack of energy and attributed it to the fact that he was getting older. Mike attended a NWSC meeting and visited in Florida for several weeks. The feelings of unease continued.

CHAPTER 15

Civil Service Korea

Some time in April 1984, the Far East District Headquarters in Seoul, Korea contacted Mike and started negotiations to get him to Korea. They were having some problems and felt that Mike could help them out. Mike had just finished a strenuous tour in Alaska. He had visited about every radar installation on the North Slope. He had visited King Salmon Cape, Newenham, Nome, Sparrauun, Tatlina, Indian Mountain, Bethel McGrath, Galena, Seward, Palmer, Whittier, Valdez, Fairbanks, Kodiak, and Denali. He had conducted upwards of twenty briefings, and participated in several pre-construction conferences. He felt a little drained.

Mike agreed to go to Korea in early May. He had not been there since 1947. His youngest daughter, son-in-law, and grandchildren were there at Osan AFB. Mike and Martha were looking forward to seeing them. Mike received travel orders for a three day stay in Honolulu and the use of a rental car while there. Mike arranged to have Martha on the same plane with him.

Back in 1980 Mike and Martha attended a NWSC meeting at Pompano Beach, Florida. They got acquainted with a Korean couple that were attending the conference also. Mike took them to the grocery store because they couldn't always get what they wanted at the restaurant. They didn't have a vehicle and Mike and Martha did. They were very pleasant people. Mike and Martha got well-acquainted with Sin Sang Pil and his family, but figured they would never see them again. They were very much mistaken.

Mike and Martha went through it all again. Mike was feeling better, but still couldn't fathom what was wrong with him. The shot of adrenaline he usually experienced when an unusual assignment was in progress had him keyed up. They flew to Honolulu, Hawaii and their vehicle was waiting at the airport. They drove to the Mauna Loa Hotel. Mike called one of his cousins who was in the Air Force and arranged for dinner for three at the club. The next morning, Mike went to the division office and arranged for the briefing. That afternoon Martha took a tour bus and went sight-seeing. Mike was well acquainted with the division safety officer and everything worked out as planned.

Mike learned that several construction projects were underway. They were making major changes in the F-16 hanger space that made the aircraft more secure. In effect, they hardened the hangers. They were constructing huge storage tanks for aviation gasoline. Much of the work was going on at the demilitarized zone. They were building some underground maintenance facilities. The Corps was also attempting to locate tunnels under the DMZ. Mike was glad to get back to that part of the world.

They left for Korea. On the way to Kimpo, Mike recalled the critical days in 1948 when he had spent one week out of the month providing air-sea rescue for the American Air Force in Korea. He had been stationed in Japan at that time with the 3rd Emergency Rescue Squadron.

It was a tiring flight to Kempo. Mike and Martha arrived about 10:00 p.m., not knowing exactly what to expect, perhaps a vehicle and driver to get to the hotel. Imagine their surprise when they saw a large sign "WELCOME THE CASEYS TO KOREA". Mike and Martha recognized who they were and let them take over. The Sins helped them get through customs, loaded their baggage on a truck, and drove them to the Ambassador Hotel in downtown Seoul. There were flowers and food in the hotel room. It quickly became evident that Mr. and Mrs. Sin were influential, well-known people.

The kids didn't dream that Mike and Martha were in Korea. The next morning they called Jeff and Martha and they came right in to Seoul and spent the rest of the day with them seeing the sights at Seoul, a town of more than 9,000,000 people.

On Monday, Mr. Sin Sang Pil picked Mike up at the Ambassador Hotel and took him to the Corps of Engineers compound. That was the last time Mike rode to work. The way to the office went through one of the busiest parts of town and certainly the most colorful. Mike thoroughly enjoyed the one mile walk to work. Martha spent most of her time with their daughter and grandchildren at Osan AFB about 25 miles from Seoul.

Mike reported to the commander of the Far East District. The District Engineer made it very clear that he was concerned about the safety of his people and that he would back Mike to the hilt. Mike was assigned to a nice office next to Mr. Sin. Mike did not try to check out a vehicle. Mr. Sin had one assigned to him. He drove Mike anywhere he wanted to go.

Mr. Sin was a very colorful character. He was assigned to the Safety Office as Mike's assistant, but he was obviously an important person in the organization. He was formerly a colonel in the South Korean Army. He was the second ranking Korean civilian in the compound. When Mike visited a construction site with Mr. Sin, a great deal of activity occurred. Everyone started getting things in good shape and gathering around to see what Sin had to say. He served as interpreter for Mike. He could write and read English better than he could speak it.

The Corps was involved with a $265,000,000 construction program. This provided new military facilities as well as repair of the aging structures that dated back to the Korean War. Mike was involved with drill crews searching for tunnels under the DMZ.

The North Koreans were thought to have drilled seven or eight tunnels. The U.S. Army had discovered several of these and were trying to uncover more by drilling in the area where they were thought to be. Drilling is a dangerous activity and Mike was involved in their safety program. Some of the tunnels were located by accident. Others were discovered when a North Korean defector told about being involved with the program. One of the tunnels was large enough to run troops through three abreast. Do this all day and the North Koreans have an army positioned in South Korea.

The UN Command Tunnel Neutralization Team had to bore through solid granite to reach the tunnel, which extended 435 yards into South Korean territory. The North Koreans accused the South of building the tunnels. One tunnel was painted black and the North Koreans claimed it was a coal mine. The North Koreans claimed that another tunnel had been built by the UN.

One pleasant evening Mr. and Mrs. Sin invited Mike and Martha to a Korean dinner at their home. They lived in a nice part of town. They owned a large two story building. They had a business on the first floor and their living quarters were on the second floor. The business was working with aluminum, making windows, doors, downspouts, gutters, etc.

Their home was beautiful. Their entire family, consisting of two sons and two daughters, were present with Mike and Martha that made eight adults dining that night. The weather was hot and humid. But! They had several air conditioners in the home and did not spare the electricity. There was a grill glowing with hot embers and they cooked the main dish over the grill. This was a dish made of thin slices of beef cooked to individual taste and then wrapped in crisp lettuce leaves with plenty of garlic cloves and other condiments. Hot peppers were available and were used liberally. Kimchee, a dish something like sauerkraut, was already prepared and ready for consumption. The grill created a lot of heat. But! The air conditioners kept it comfortable. The Sins said that not a week went by without the family dining together. It was one of the more pleasant meals Mike could remember.

When Mike and Mr. Sin traveled together, which was often, Mr. Sin always looked for a Korean restaurant in the native section of town. He would fire up the grill and cook a truly good meal. Mike ate so much garlic that Martha said his elbows smelled like garlic.

Almost every base in Korea had good dining facilities. The Non-Commissioned Officers' Club was usually the best. The civilian workers also had a club and the food was good. There were several slot machines at these clubs. Whenever Mr. Sin and Mike conducted a safety survey or visited a project, there was always a lot of activity. Anytime the workers saw his vehicle, they really came alive. They rushed around making things look good and they made sure that everyone knew that Mr. Sin was there. Mike had no problem with getting their attention.

One weekend Mike took Martha on a tour of the DMZ. He took her places that other visitors were not permitted to go. They spent some time at Panmunjom and they visited several construction projects in the area. While there, they had lunch at a very good Non-Com Club in the area. Mike and Martha had the dubious honor of stepping over into Communist North Korea. They took pictures of the tree where the American soldiers were killed prior to their visit.

One trip Mike really enjoyed was a chopper flight from Seoul to Pusan. It was a lovely day, visibility was unlimited and there was very little turbulence. Korea is a very lovely country when seen from the air.

Mike was beginning to tire easily and had trouble using his left hand. He knew things were not right as far as his health was concerned, but since he had three physical exams in the last three years, he again put it in the back of his mind. Subconsciously, he just didn't want to think about it.

On their days off Mike and Martha did a lot of shopping. They bought eight or ten windbreaker-type jackets. Mike had a tailor make a beautiful tweed sports jacket. Martha bought a lovely silk dress and a raincoat that really looked good on her. Shopping in Korea was a lot of fun. At most of the smaller shops, haggling over the price was commonplace. Mike and Martha bought several pair of running shoes.

Mike and Martha left for home on August 3, 1984. They flew to Tokyo and caught a non-stop flight from Tokyo to New York. It was a back-breaking flight. Fourteen hours non-stop. They arrived in Huntington worn out and glad to be home.

CHAPTER 16

Retirement

In 1985 Mike decided to run for his dad's old house seat in the State Legislature. He even got so far as to fill out the necessary papers and was ready to pay the filing fee. He lined up enough support to start a campaign. He decided to call it quits. He didn't have enough of his vitality left to mount a winning campaign. Martha could see a big change in him. The aggressive, ambitious, outgoing man she was used to just wasn't around anymore.

In May 1986, Mike was contacted by the Chief of Safety in Frankfurt. He wanted Mike to come to Turkey and manage the safety program for the summer. The duties involved NATO bases in Turkey where many construction projects were in progress. They had reason to believe Mike could do a job for them. It was a done deal. Mike was to fly to Frankfurt by commercial air and catch an Air Force flight to Incirlik. Mike was all set to go. The catch was that he had to take a physical. This time Mike went to see a neurologist, who told him he had some symptoms of Parkinson's disease. The doctor would not make a definite diagnosis, but prescribed some pills for Mike to take. The doctor recommended that Mike not make the trip to Turkey. He was advised that he should not be running all over the world. This was the ignominious end to the Turkey venture. Mike could read the handwriting on the wall. He didn't have to get hit in the head with a sledge hammer. Never again would he get off an airplane with his luggage and a brain full of knowledge about how to make the workplace better and protect life and property. Mike decided not to go and suffered another bitter disappointment.

In late 1987 the neurologist made a definite decision that Mike, Jr. Casey had Parkinson's disease. Mike didn't know which way to turn. He had been going around and around for several months. One of the first things he did was get a second opinion. Mike and Martha traveled to Cleveland, Ohio and Mike was examined by experts in this field. The same diagnosis was arrived at—Parkinson's disease. By this time Mike was really feeling very bad. He stayed home most of the time. He had trouble reading and was uncomfortable working crossword puzzles, two activities he really enjoyed.

Mike and Martha joined a Parkinson's support group and attended meetings once a month. Apparently, Parkinson's is a disease in which a person does not lose the power to reason. They sit back and watch themselves slowly become incapacitated and know that there is very little they can do about it. It is difficult for a person to adjust to the fact that there is no known cure. It becomes harder and harder to write a letter or sign a check.

Martha had to do all the driving and shopping. Mike was so busy feeling sorry for himself that he didn't know what Martha was going through. She was just as bewildered with the turn of events as Mike was. Mike and Martha designed a new home and hired a contractor to build it. The house was designed with the handicapped in mind. It had a ramp from the garage to the kitchen and wide doors and wide hallways. There were grab bars in the bath and shower and other features designed for barrier-free living.

They lived in the house for almost four years. The lawn care and other maintenance got to be a chore and they moved into a retirement community, which offered special medical care if needed. After five or six years as a virtual recluse, Mike started a few activities. Mike began to get involved in a few card games. He organized his slides and put on a few shows.

Mike decided that he wanted to do something for his daughters that they would always remember. He decided to take his family on a trip to Israel. He realized that it would be expensive and would play havoc with his waning supply of energy. Mike had seen the country as few others ever had the chance to see it. He had been the length and breadth of the Sinai Desert and traveled extensively throughout the country. Mike had already visited the country twice. When he arrived in Israel the first time, he felt like he had never been anywhere before. He admired the people and felt that the aggressive defense of their little country was justified.

Israel was a dangerous place where terrorism and violence was concerned. However, it was one of the safest places as far as other felony crimes. Rape, murder, and other crimes of violence were pretty well controlled. Justice was swift and certain when crime was committed. Women hitched-hiked all over the country day and night and were never at great risk. Sex crimes were at a minimum. All in all it was a good place to be. All of the youth were required to spend some time in the service to the country. Some chose the army, others opted for service at a kibbutz. Many of today's leaders were raised on these collective farms.

Policymakers and leaders in the army were not selected because of their connections or desire to command. A ribbon clerk could be a high ranking officer if he were a true leader. Sabras or people who were born in Israel are some of the best fighters in the world. They manage to select and train the best pilots in existence. Some were still in their late teens.

Mike had always admired Menachem Begin. He was not overly impressive physically, but pound for pound he was probably one of the more dangerous people in the Middle East. It would be difficult to ascertain how many men he had wasted in his younger years when the Jews were fighting for their very existence.

The girls came home to get ready for the trip. Sandy broke her foot, but still made the trip. She borrowed one of her grandfather's canes. Mike advised Sandy to leave all of her Air Force identification at home. There was too much hostage-taking for an individual to be complacent. The girls made several purchases and paid too much for them. Mike bought some of the same articles and haggled the merchants down to about one third of what the girls paid.

While in Amman, Jordan, a near tragedy occurred. Mike, Sandy, and Martha were traveling up an escalator, when a lady decided she wanted to go down the up escalator. The woman sandbagged them about halfway down the escalator. Four or five people fell including Mike, Sandy, and Martha. The escalator continued to run until Sandy shouted in her best parade ground voice "Stop the damned escalator!" The equipment stopped. Mike's hand was bleeding and Sandy's skirt was torn.

From this point on it was all uphill for Mike. He was getting progressively worse, but could do nothing about it. They arrived home still very excited about the trip. Mike just moped around. He could barely manage to mow the grass. He stopped all his old activities. Tension was beginning to mount. Martha was just as confused as Mike. They went to Florida and took a cruise ship.

Mike does not hesitate to be realistic where his illness is concerned, he has lost much of his aggressiveness and does not speak out like he once did when he was full of confidence in himself. He doesn't defend his position on controversial subjects as ardently as he used to. He would probably walk away from a confrontation if it required physical activity.

One of the more noticeable symptoms of PD is referred to as the Parkinson's Mask. It is a constant scowl or frown. Mike lets his friends know that he does not have control over his facial expressions to any great extent, just as he feels that most of the time he has no control over his life.

Mike had great regard for the flying equipment he had accumulated over the years. He kept it until August 1995. He has his A-2 Jacket. He gave everything else back to the Air Force. It was donated to the Wright-Patterson Air Force Museum. The fur-lined boots were in excellent shape. Mike only wore them for about three or four months while he was flying the Hump to China.

Many of the flight crews wore electrical flying suits, but Mike preferred the fur. The fur-lined suits never shorted out. The items Mike donated are listed in the letter attached to this writing. It may be interesting how Mike obtained some of the equipment. He traded a case of beer for the binoculars. A British flier salvaged them from a C-47 that had made a crash landing and traded them to Mike. The leather B-3 Jacket was issue and Mike had no trouble getting it through customs. The headsets were salvaged from an old crash boat in Japan. It was common for collectors to pay hundreds of dollars for some of this equipment, but it is now back where it belongs. If any of Mike's grandchildren would like to see his equipment, they can use the catalogue number in the attached letter and have access to the items.

Mike and Martha moved into a retirement community on Route 52, just across the interstate from Huntington. Woodlands has a very good health unit. It also has a fine library. Most of the manuscript for this book was written in long hand at the library. Mike's eldest daughter worked long hours typing the material. Mike will never go around and around again with the same verve and elan he displayed in the early years. However, he intends to continue to go around and around for some time to come. NUFF SAID!

Casey has many awards and other documents, indicating outstanding performance. A few of these papers are contained in this book.

—Letter from Colonel John Devans, reference retirement.

—Letter of Appreciation, Colonel Higman

—Letter of Appreciation, Colonel Roush

—Letter from the United States Air Force Museum

—Letter of Appreciation, Colonel Devens

—Letter of Appreciation from Colonel John Lillibridge, reference the Safeguard assignment.

—Letter of Appreciation from Colonel William E. Lee, reference the assignment to Israel.

—Letter from General Starnes, reference Buffalo Creek disaster.

—Letter from Cordell Hull, the Seceretary of State under Roosevelt, reference Casey's release from the Canadian Armed Forces.

DEPARTMENT OF THE ARMY
HUNTINGTON DISTRICT, CORPS OF ENGINEERS
502 EIGHTH STREET
HUNTINGTON, WEST VIRGINIA 25701

REPLY TO
ATTENTION OF:

December 30, 1982

Mr. Robert L. Casey
Safety Office

Dear Bob:

I want to take this opportunity to express my thanks for your 16 years of outstanding service with the Huntington District.

Your active participation in the execution and direction of the Safety Program has resulted in an enviable position for the District. Your efforts have resulted in many recognitions, including two Division Safety Awards. Your leadership in safety training activities for field and office personnel has been of highest caliber and has produced excellent results. Your surveillance over both the District personnel as well as contractor operations has materially reduced accidents and has resulted in thousands of employees being more safety conscious in their present and future work.

I am extremely well pleased with your administration of the Health Unit. The Health Unit continues to operate at a high level and provides both office and field training for District personnel.

During your 16 years of service with the Huntington District you have compiled an enviable record of successful completion of many and varied assignments. Your participation on many Boards of Investigation, your service on the Safeguard missile program as well as the Siani Construction Program and your leadership in the National Water Safety Congress are indicative of the far-reaching services which you so willingly executed and which has reflected much credit on the Huntington District office. The four outstanding performance ratings testify to your continued high standards of service.

I know the employees of the District join with me in wishing you a most rewarding retirement, and to let you know that we will miss your dedication to the Corps' Safety Program.

Sincerely,

John W. Devens
Colonel, Corps of Engineers
District Engineer

DEPARTMENT OF THE ARMY
HUNTINGTON DISTRICT, CORPS OF ENGINEERS
P. O. BOX 2127
HUNTINGTON, WEST VIRGINIA 25721

REPLY TO
ATTENTION OF:

ORHEA

24 April 1981

SUBJECT: Letter of Appreciation

Mr. Robert L. Casey
Safety Office

The Division Safety Award is a significant accomplishment for this District. The award means much more than recognition. It represents great savings of time, money and most important - an accident-free workforce. It had to be a team effort. You led the team and very successfully. I am aware of your constant work in accident prevention. It has been a job well done. Congratulations!

JAMES H. HIGMAN
Colonel, Corps of Engineers
District Engineer

DEPARTMENT OF THE ARMY
HUNTINGTON DISTRICT, CORPS OF ENGINEERS
P. O. BOX 2127
HUNTINGTON, WEST VIRGINIA 25721

IN REPLY REFER TO:

ORHEA

8 November 1968

SUBJECT: Letter of Appreciation

Mr. Robert L. Casey
Acting Chief, Safety Office

1. Please accept my sincere appreciation for your outstanding assistance in the preparations for and execution of the recent dedication activities. The requirement to superimpose planning for the Fishtrap Reservoir dedica-

tion upon the dedication plans for Grayson Reservoir required maximum effort and flexibility. The accomplishment of these tasks required close cooperation by many District Office and field elements and outside organizations such as local, county, and state law enforcement officials; rescue team directors; and Red Cross and ambulance teams.

2. Your thoroughness, knowledge and devotion to duty contributed significantly to the success of both dedications. You aggressively accomplished your assignments relative to establishing helipads, providing emergency safety coordination and keeping the project officers informed of safety-related events.

3. You may be justly proud of the part you had in these operations, for your performance reflects great credit to you, the Huntington District, and the Corps of Engineers.

MAURICE D. ROUSH
Colonel, Corps of Engineers
District Engineer

CF:
Personnel Folder, Mr. Casey

DEPARTMENT OF THE AIR FORCE
UNITED STATES AIR FORCE MUSEUM
1100 SPAATZ ST
WRIGHT-PATTERSON AIR FORCE BASE, OHIO 45433-7102

30 Oct 95

USAFM/DR
Bldg 489
1100 Spaatz St
Wright-Patterson AFB OH 45433-7102

Lieutenant Colonel Robert L. Casey, USAF Retired
6218 Brevda Ct
Huntington WV 25705

Dear Colonel Casey

With genuine appreciation, I accept on behalf of the United States Air Force and the USAF Museum your flying jacket, boots, goggles, clothing and binouculars as cataloged on the attached list They have been added to the USAF Museum collection and permanently recorded as a donation under Accession Number 1995-096.

Your items have been placed in our study collection, since less than 20 percent of our items are on public display. The remaining artifacts are carefully preserved, and serve as a resource for researchers and historians and for future exhibits.

Providing exhibits of historic items for today's visitors is only part of our responsibility. Equally important are the acquisition and preservation of items for tomorrow.

Through your gift you have contributed significantly to the preservation of our history, our Air Force heritage and tradition, and the memory of those who led the way.

Thank you very much for your generosity and patronage.

Sincerely

RICHARD L. UPPSTROM
Director

USAF MUSEUM GIFT ACCESSION

DONOR Lt. Col. Robert L. Casey
USAF, Retired
6218 Brevda Ct.
Huntington WV 25705

ITEM RECEIVED 05 Sep 95

ITEM NUMBER 1995-096

MUSEUM CATALOG NUMBER	ITEM DESCRIPTION
1995-096-001	Jacket, Flying, Winter, Leather, B-3
1995-096-002	Binoculars, M3, W/O Case
1995-096-003	Boots, Flying, A-6A
1995-096-004	Radio Receiver, ANB-H-1
1995-096-005	Can, Cutfit Anti-Dimming, MK-1
1995-096-006	Goggles, Flying
1995-096-007	Underwear
1995-096-008	Hood, Knit

ORHEA (21 May 1982) (25/COR-43) 1st Ind
SUBJECT: Letter of Appreciation

DA, Huntington District, Corps of Engineers, 502 Eighth Street, Huntington, West Virginia 25701 14 JUL 1982

TO: Robert L. Casey, Chief, Safety Office

1. I am advised that Colonel Higman was apprised of your accomplishment in the Sinai, and he provided me with Colonel Lee's letter for my information.

2. It is gratifying to hear from a Commander that the quality of service rendered is of the highest caliber. I understand you have also distinguished yourself in the Ohio River Division as well as this District. Your professionalism and expertise were obviously considered in your selection for the TDY assignment.

3. My appreciation is offered for your commendable performance with the Sinai Construction Management Office, and I will rely on you to continue the Huntington District Safety Program as the Division leader.

John W. Devens
JOHN W. DEVENS
Colonel, Corps of Engineers
Commanding

DEPARTMENT OF THE ARMY
SAFEGUARD AREA OFFICE, CORPS OF ENGINEERS
PO BOX 631
LANGDON, NORTH DAKOTA 58249

MROSA-G

15 September 1970

SUBJECT: Letter of Appreciation

THRU: District Engineer
Huntington District
P.O. Box 2127
Huntington, West Virginia 25721

TO: Robert L. Casey
Safety Engineer
Huntington District
P.O. Box 2127
Huntington, West Virginia 25721

1. You were assigned to the Grand Forks SAFEGUARD Area Office for about six weeks starting in early August and ending last week. During that time you displayed a willingness to work, a spirit of helpfulness and cooperation, and a high degree of technical competence in the safety field. Your primary mission was to assist the Area Office in setting up and getting off to a good start on safety in construction. You discharged your duties with zest and in a professional manner.

2. I acknowledge and appreciate your contribution to the mission of this office.

John L. Lillibridge
JOHN L. LILLIBRIDGE
COLONEL, CE
Area Engineer

DEPARTMENT OF THE ARMY
U.S. ARMY CORPS OF ENGINEERS
SINAI CONSTRUCTION MANAGEMENT OFFICE
APO NEW YORK 09672

REPLY TO
ATTENTION OF:

SCMZA

2 1 MAY 1982

SUBJECT: Letter of Appreciation

Colonel James H. Higman
US Army Engineer District, Huntington
P.O. Box 2127
Huntington, West Virginia 25721

Dear Colonel Higman:

I wish to express my appreciation to you for your support of the mission of the Sinai Construction Management Office. The TDY support provided by your Safety Engineer, Mr. Robert L. Casey, was extremely valuable.

Mr. Casey's role in this project assisted us in meeting a near-impossible schedule. SCMO was tasked to construct two base camps and 42 outposts in the Sinai for the Multinational Force and Observers, in accordance with the Camp David Peace Treaty. This task involved a peak employment of over 900 persons, working at sites extending the entire length of the Sinai Peninsula. Mr. Casey's continued and conscientious appraisal of construction safety conditions, and his recommendations for implementing safe working environments, were largely responsible for this office's respectable safety record. The fact that SCMO was able to meet its schedule with a minimum loss of man-hours and resources clearly reflects on Mr. Casey's technical skills. We are indeed grateful for his assistance.

Sincerely,

WILLIAM E. LEE, JR.
COL, CE
Commanding

DEPARTMENT OF THE ARMY
OHIO RIVER DIVISION, CORPS OF ENGINEERS
P. O. BOX 1159
CINCINNATI, OHIO 45201

ORDDE 14 July 1972

Mr. Robert L. Casey
Safety Office
Huntington District

Dear Mr. Casey:

A sudden failure at an impoundment on Buffalo Creek in Logan County, West Virginia, on the morning of 26 February 1972 released thousands of gallons of water into a narrow, populated valley causing widespread devastation along a seventeen mile path, resulting in the death of one hundred and eighteen residents. Roads and utilities in the area were washed away. When the catastrophic deluge ended, there were tons of debris blocking the channel of the creek and spread over the valley. Health hazards abounded in the ruined area. Human bodies were buried in wreckage. Hundreds of damaged homes had to be razed and over one thousand demolished automobiles had to be removed. Hundreds of people were rendered homeless.

On 27 February the Huntington District was assigned the mission of debris removal and cleanup. However, local rains around the District Office and through the State of West Virginia caused numerous road blockages by high water and land slides. Travel to the disaster site by normal routes was impossible. You volunteered to accompany the Office-In-Charge to the disaster scene. Your knowledge of alternate roads, gained while serving with the West Virginia State Police, enabled the Office-In-Charge to bring his first contingent on the scene with minimal delay.

In those early hours of the tragedy, there was heavy traffic and confusion. The State Police, hindered by limited manpower, was trying to bring order out of confusion. Access to the disaster area was restricted and road movement was difficult. Your experience and knowledge of key State Police personnel, coupled with considerable tact and persuasion, gained for Corps and contractor personnel the carte blanche authority to come and go as the necessity of the mission dictated.

You willingly and unhesitatingly assisted in performing a multitude of other necessary tasks in connection with safety and health requirements of the situation.

ORDDE 14 July 1972

Mr. Robert L. Casey

In your role as Safety Officer you conducted various safety meetings for National Guard, Corps and contractor personnel. After your return to the District Office, you returned often to the disaster site to assist and advise Corps supervisors in safety matters. The accident-free record of the operations is due, in part, to your diligence.

Your long hours of work, resourcefulness, and devotion to duty is deeply appreciated.

Sincerely yours,

W. L. Starnes

W. L. STARNES
Major General, USA
Division Engineer

DEPARTMENT OF STATE
WASHINGTON

In reply refer to
Eu 842.2221/161

September 6, 1941

My dear Senator Rosier:

I have received your letter of August 27, 1941 with regard to the desire of Mr. Mike Casey of Huntington, West Virginia, to ascertain the whereabouts of his son, Private R. L. Casey, who is serving in the Canadian Army. This information is desired in connection with efforts being made to obtain Private Casey's release from the Canadian forces.

Confirming the information which was telephoned to your secretary on August 26 you are informed that the Department has ascertained from the Canadian Legation at Washington that Private Casey's address is No. 3-A, District Depot, (SDG Wing), Kingston, Ontario. It was also learned that the Legation

The Honorable

Joseph Rosier,

United States Senate.

had recommended to the appropriate authorities in Ottawa on August 16 that Private Casey be released from the Canadian armed forces. This information was conveyed to Mr. Mike Casey by the Legation in a letter dated August 22, and it was suggested at the same time that Mr. Casey communicate with his son with a view to having Private Casey submit, in accordance with the usual practice, a request that he be released.

Sincerely yours,

About the Author

Casey retired from the air Force reserves about this time. 17 May 1981. He was 60 years old at this point in his career.

Bob Casey believes that it is every man's right to be a citizen soldier. He has always supported the reserve program with great enthusiasm. His total service time amounted to about 33 years. He enlisted in the old Army Air Corps as a private on July 8, 1942, and was awarded a reserve commission in 1948. During the time he was a State Trooper, he also served as an active reservist. Bob was discriminated against for years because of his reserve activities.

The right of a state employee to be awarded military leave for training purposes has been established for years. The DPS refused to comply with state law on several occasions. Bob used his personal leave time to meet the Air Force requirements. Bob was the only Reserve Officer in the Department at that time. Bob remembers the days when he found it necessary to appear before the Military Affairs Committee and relate his experiences as a reservist in the Department. The members of the committee couldn't believe that the Department of Public Safety was not in compliance with state law, especially the chairman of the group who was a Lieutenant Colonel in his own right and trained with the reserves. Legislation was passed that the top brass resented to the end. It was direct and to the point. The Department of Public Safety will grant up to 30 days a year in military leave to an Active Reservist who requests it. Bob Casey got military leave and some flack with it. It is difficult to be efficient and effective when you feel that you are not getting a fair shake. This situation does not exist today. Reservists get a fair shake and are respected members of the DPS. Bob believes that the turning point was reached when General Mooney became Superintendent.

For what it is worth, Bob is a graduate of the Command and Staff College with two weeks in residence at the Air University in Montgomery, Alabama and one year of correspondence. He is a graduate of the Armed Forces Industrial College and the U.S. Army Safety Management Course.

Several Years ago, the Department sponsored legislation to give Troopers who were veterans credit for their military time. The limit was five years. Bob Casey did not receive these credits. He was told since he was drawing a partial disability pension he was not eligible for the military credits. Bob feels that since he was on active duty during the war and his service was honorable, he should be eligible. He is not mad. The members of the Department used to speculate as to how many checks Bob would receive after retirement. He gets several checks and lives comfortably. As long as they are legitimate, it should not be a matter of concern.

Bob still remembers the day he was named as Commander of two Reserve flights in Princeton—more exposure, more responsibility and more prestige. Something leaked and Bob was scheduled for transfer. He volunteered to go to Welch in order to stay in the general vicinity and command his flights. The Department sent him about 200 miles away to Philippi. He gave up his command. Recently, Bob wrote the Retirement Board for a hearing. However, he is almost 78 years old and will not win very many longevity contests. 'Nuff said.

Even after all of this, Bob still thinks that the DPS is the best law enforcement agency in the United States. He is pleased and proud he and his brother served with the Department.

Bob Casey and the Women in his life
L-R Marty, Martha and Sandy